Atomic Absorption and Plasma Spectroscopy

Analytical Chemistry by Open Learning
Second Edition

Author:
JOHN R. DEAN
University of Northumbria at Newcastle

Editor:
DAVID J. ANDO
University of Greenwich

Author of First Edition:
ED METCALFE
University of Greenwich

Published on behalf of ACOL (University of Greenwich)
by
JOHN WILEY & SONS
Chichester • New York • Weinheim • Brisbane • Singapore • Toronto

Copyright © 1997 University of Greenwich, UK
Published by John Wiley & Sons, Ltd,
          Baffins Lane, Chichester,
          West Sussex PO19 1UD, England

          National 01243 779777
          International (+44) 1243 779777

          e-mail (for orders and customer service enquiries): cs-books@wiley.co.uk
          Visit our Home Page on http://www.wiley.co.uk
                    or http://www.wiley.com

Other Wiley Editorial Offices

John Wiley & Sons, Inc., 605 Third Avenue,
New York, NY 10158-0012, USA

VCH Verlagsgesellschaft mbH, Pappelallee 3,
D-69469 Weinheim, Germany

Jacaranda Wiley Ltd, 33 Park Road, Milton,
Queensland 4064, Australia

John Wiley & Sons (Asia) Pte Ltd, 2 Clementi Loop #02-01,
Jin Xing Distripark, Singapore 129809

John Wiley & Sons (Canada) Ltd, 22 Worcester Road,
Rexdale, Ontario M9W 1L1, Canada

Library of Congress Cataloging-in-Publication Data

Dean, John R.
    Atomic absorption and plasma spectroscopy / author, John R. Dean ;
    editor, David J. Ando. — 2nd ed.
        p.   cm. — (Analytical Chemistry by Open Learning)
    Rev. ed. of: Metcalfe, Ed. Atomic absorption and emission
spectroscopy. c1987.
    Includes bibliographical references (p.   —   ) and index.
    ISBN 0-471-97254-1 (cloth : alk. paper). — ISBN 0-471-97255-X
(pbk. : alk. paper)
    1. Atomic spectroscopy — programmed instruction.   2. Plasma
spectroscopy — programmed instruction.   3. Chemistry, Analytic-
Programmed instruction.   I. Ando, D. J. (David J.)   II. Metcalfe,
Ed. Atomic absorption and emission spectroscopy.   III. Title.
IV. Series: Analytical Chemistry by Open Learning (Series)
QD96.A8D4   1997
543'.0858 — dc21   96–47643
                                                                  CIP

British Library Cataloguing in Publication Data

A catalogue record for this book is available from the British Library

ISBN 0 471 97254 1 (cloth)
ISBN 0 471 97255 X (paper)

Typeset in 11/13pt Times by Mackreth Media Services, Hemel Hempstead, Herts
Printed and bound in Great Britain by Biddles Ltd, Guildford, Surrey
This book is printed on acid-free paper responsibly manufactured from sustainable forestation,
for which at least two trees are planted for each one used for paper production.

To Lynne, Samuel and Naomi

THE UNIVERSITY OF GREENWICH
ACOL PROJECT

This series of easy-to-read books has been written by some of the foremost lecturers in Analytical Chemistry in the United Kingdom. These books are designed for training, continuing education and updating of all technical staff concerned with Analytical Chemistry.

These books are for those interested in Analytical Chemistry and instrumental techniques who wish to study in a more flexible way than traditional institute attendance, or to augment such attendance.

ACOL also supply a range of training packages which contain computer software together with the relevant ACOL book(s). The software teaches competence in the laboratory by providing experience of decision making in such an environment, often based on the simulation of instrumental output, while the books cover the requisite underpinning knowledge.

The Royal Society of Chemistry uses ACOL material to run regular series of courses based on distance learning and regular workshops.

Further information on all ACOL materials and courses may be obtained from:

The ACOL–BIOTOL Office, University of Greenwich, Unit 42, Butterly Avenue, Dartford Trade Park, Dartford, DA1 1JG. Tel: 0181-331-7533, Fax: 0181-331-9672.

How to Use an Open Learning Book

Open Learning books are designed as a convenient and flexible way of studying for people who, for a variety of reasons, cannot use conventional education courses. You will learn from this book the principles of one subject in Analytical Chemistry, but only by putting this knowledge into practice, under professional supervision, will you gain a full understanding of the analytical techniques described.

To achieve the full benefit from an open learning text you need to carefully plan your place and time of study.

● Find the most suitable place to study where you can work without disturbance.

● If you have a tutor supervising your study discuss with this person the date by which you should have completed the text.

● Some people study perfectly well in irregular bursts; however, most students find that setting aside a certain number of hours each day is the most satisfactory method. It is for you to decide which pattern of study suits you best.

● If you decide to study for several hours at once, take short breaks of five or ten minutes every half hour or so. You will find that this method maintains a higher overall level of concentration.

Before you begin a detailed reading of this book, familiarise yourself with the general layout of the material. Have a look at the course contents list at the front of the book and flip through the pages to get a general impression of the way the subject is dealt with. You will find that there is space on the pages to make comments alongside the text as you study — your own notes for highlighting points that you feel are particularly important. Indicate in the margin the points you would like to discuss further with a tutor or fellow student. When you come to revise, these personal study notes will be very useful.

Π   When you find a paragraph in the text marked with a symbol such as is shown here, this is where you get involved. At this point you are directed to do certain things, e.g. draw graphs, answer questions, perform calculations, etc. Do make an attempt at these activities. If necessary, cover the succeeding response with a piece of paper until you are ready to read on. This is an opportunity for you to learn by participating in the subject, and although the text continues by discussing your response, there is no better way to learn than by working things out for yourself.

We have introduced self-assessment questions (SAQs) at appropriate places in the text. These SAQs provide you with a way of finding out if you understand what you have just been studying. There is space on the page for your answer and for any comments you want to add after reading the author's response. You will find the author's response to each SAQ at the end of the book. Compare what you have written with the response provided and read the discussion and advice.

At intervals in the text you will find a Summary and a list of Objectives. The Summary will emphasise the important points covered by the material you have just read, while the Objectives will give you a checklist of tasks you should then be able to achieve.

You can revise the book, perhaps for a formal examination, by re-reading the Summary and the Objectives, and by working through some of the SAQs. This should quickly alert you to areas of the text that need further study.

At the end of this book you will find, for reference, lists of commonly used scientific symbols and values, units of measurement, and also a periodic table.

Contents

Study Guide

The first Atomic Absorption and Emission Spectroscopy Unit in the ACOL series was written by Ed Metcalfe. This successor to that Unit has been written by John R. Dean, who is currently Reader in Applied Chemistry at the University of Northumbria at Newcastle.

This present text has been arranged to provide you with a working knowledge of modern atomic absorption and plasma spectroscopy. The alteration in title from the first Unit has allowed new developments in plasma-based detection systems to be incorporated, e.g. mass spectrometry.

It will be assumed that you have an understanding of chemistry equivalent to that of a student who has completed one year of study in Higher Education. Alternatively, some working knowledge of atomic spectroscopic techniques would be beneficial.

As this text covers only the essentials of atomic absorption and plasma spectroscopy, you are bound to find that there are some topics that you would like to study in more detail than is given in the text. Suitable books that you could use as a starting point are listed either in the Bibliography or in the Further Information chapter at the back of the book (Section 8.2; Additional Reading Material). For the most up-to-date information on aspects of atomic absorption and plasma spectroscopy you are directed towards articles in relevant journals (suitable journals are also included in Section 8.2).

You will find that the best way to learn is through hands-on experience. As with most analytical techniques, knowledge comes through time spent in front of the appropriate instrument. Through the use of selected experiments and the material provided in this Unit you should become suitably proficient in a relatively short time-scale.

Supporting Practical Work

1. GENERAL CONSIDERATIONS

The range of atomic spectroscopy instrumentation that is available will vary widely from one laboratory to another. The experiments that follow are designed to illustrate basic procedures and hence they should be considered as only being introductory in nature.

2. AIMS

(a) To provide experience in the preparation and pre-treatment of samples in a form suitable for analysis.

(b) To provide experience of operating and optimising appropriate instrumentation.

(c) To select a suitable working range for a given element, and construct and use a calibration graph.

(d) To demonstrate the interferences which can arise, and the methods used for combating these interferences.

(e) To demonstrate the use of other methods of calibration, e.g. standard additions and isotope dilution analysis.

3. SUGGESTED EXPERIMENTS

Flame Atomic Absorption Spectroscopy

(a) Effect of phosphate addition to the atomic absorption of the calcium atom; use of lanthanum, strontium or ethylenediaminetetraacetic acid (EDTA) to alleviate the problem.

(b) The determination of iron in beer by using ammonium pyrolidine dithiocarbamate (APDC)/methyl isobutyl ketone (MIBK) extraction.

Graphite Furnace Atomic Absorption Spectroscopy

(c) Optimisation of ashing and atomisation for cadmium and copper.

(d) Determination of arsenic by using nickel chloride as a matrix modifier.

Inductively Coupled Plasma-Atomic Emission Spectroscopy

(e) Optimisation of inductively coupled plasma (ICP) operating parameters, effect of gas flow rates, viewing height and radio frequency (RF) power on the $Ca(I)$ and $Ca(II)$ signals.

(f) Comparison of linear dynamic range, sensitivity and detection limits with flame atomic absorption spectroscopy. If a polychromator spectrometer is available, the generation of data simultaneously by using mixed element standards.

Inductively Coupled Plasma-Mass Spectrometry

(g) Study of potential interferences, e.g. molecular interferences.

(h) Use of isotope dilution analysis of lead in wine.

Other Experiments

(i) Analysis of trace elements in (certified) reference materials. This may require the digestion of a solid sample by using concentrated acids. Compare the results obtained by using a calibration graph with those obtained by the method of standard additions.

Bibliography

There are numerous books available which detail the theory, instrumentation and applications of atomic absorption and plasma spectroscopic techniques (see Section 8.2). However, for alternative introductions to the area of atomic spectroscopy you might wish to consult the following general texts:

H.H. Willard, L.L. Meritt, Jr, J.A. Dean and F.A. Settle, Jr, *Instrumental Methods of Analysis*, 7th Edn, Wadsworth Publishing Company, Belmont, CA, 1988.

D.A. Skoog and J.L. Leary, *Principles of Instrumental Analysis*, 4th Edn, Saunders College Publishing, Orlando, FL, 1992.

C. Vandecasteele and C.B. Block, *Modern Methods of Trace Element Determination*, Wiley, Chichester, 1993.

A.G. Howard and P.J. Statham, *Inorganic Trace Analysis: Philosophy and Practice*, Wiley, Chichester, 1993.

G. Christian, *Analytical Chemistry*, 5th Edn, Wiley, Chichester, 1994.

Acknowledgements

Figure 1.5 is redrawn from the National Bureau of Standards Certificate of Analysis, *Standard Reference Material 1572: Citrus Leaves*, published by The National Bureau of Standards, Washington, DC (1982). Permission has been requested.

Figure 5.3 is redrawn from J. Davies, J.R. Dean and R.D. Snook, *Analyst*, **110**, 535 (1985) and is reproduced with the permission of The Royal Society of Chemistry.

Figures 6.2a, 6.2b, 6.2c and 6.3c–6.3e are taken from various applications literature published by VG Elemental and are reproduced with permission.

Figure 6.2f is redrawn from K.E. Jarvis, A.L. Gray and R.S. Houk, *The Handbook of Inductively Coupled Plasma Mass Spectrometry*, Blackie Academic and Professional, Glasgow, 1992, p. 45, and is reproduced with permission.

Figures 6.2g and 6.3f are taken from various applications literature published by Micromass. Permission has been requested.

1. Methodology in Trace Element Analysis

The determination of trace elements is necessitated by their importance in the chemical and pharmaceutical industries and in the environment. While the needs of these areas may all be different they all have a requirement to measure trace elements. This book considers the techniques that are the main backbone of analytical atomic spectroscopy. Although other techniques, such as X-ray fluorescence and ion chromatography, are utilised, it is often atomic absorption spectroscopy (flame and graphite furnace) and inductively coupled plasmas with either atomic emission or mass spectrometric detection that are the major techniques used today. This book will outline the basic principles of these techniques and suggest instrument modifications that are currently used.

1.1 INTRODUCTION

The determination of elements is essential to our own existence. The food that we eat, the water we drink, the computer that I am using to write this book, the medicine that I obtain if I am ill, the plastic chair that I am sitting on, the car I travel to work in — all of these things contain elements. Quite often, it is the amounts of the elements present that govern whether the product, material or my health is suitable for the job it has to do. If any of these fail, it is necessary to understand what caused the breakdown, or better still, to be aware before any problems arise what it is which governs their performance. Therefore, the measurements of elements (metals and metalloids), over a range of concentrations from percentage levels through to trace (parts per million) and ultratrace (parts per billion and below) levels, are the reasons for which the techniques that comprise atomic spectroscopy are used.

Table 1.1 Average elemental composition of a 70 kg adult human body[a]

Element (symbol)	Mass (g)	Concentration ($mg\,kg^{-1}$)	Year of discovery as an essential element and some functions and deficiency aspects
Oxygen (O)	45 500	650 000	
Carbon (C)	12 600	180 000	
Hydrogen (H)	7000	100 000	
Nitrogen (N)	2100	30 000	
Calcium (Ca)	1050	15 000	
Phosphorus (P)	700	10 000	
Sulfur (S)	175	2500	
Potassium (K)	140	2000	
Chlorine (Cl)	105	1500	
Sodium (Na)	105	1500	
Magnesium (Mg)	35	500	
Iron (Fe)	4.2	60	17th century; involved in oxygen and electron transport. Deficiency results in anaemia
Zinc (Zn)	2.3	33	1896; constituent of a large number of enzymes. Deficiency leads to growth retardation, sexual immaturity, skin lesions, etc.
Silicon (Si)	1.4	20	1972; connected with calcification (bones)
Rubidium (Rb)[b]	1.1	16	
Fluorine (F)	0.8	11	1971; essential for structure of teeth. Deficiency leads to increased incidence of caries; toxic at high concentrations
Zirconium (Zr)[b]	0.3	4	
Bromine (Br)[c]	0.2	3	
Strontium (Sr)[b]	0.14	2	
Copper (Cu)	0.11	2	1928; linked to oxidative enzymes; interacts with iron. Deficiency results in anaemia

Element			
Aluminium (Al)[b]	0.10	1	
Lead (Pb)[c]	0.08	1	
Antimony (Sb)[b]	0.07	1	
Cadmium (Cd)[c]	0.03	0.4	1977
Tin (Sn)[c]	0.03	0.4	1970; believed to be essential for growth
Iodine (I)	0.03	0.4	1820; constituent of thyroid hormones. Deficiency results in goitre and depression of thyroid function; excess leads to thyrotoxicosis
Manganese (Mn)	0.02	0.3	1931; participates in mucopolysaccharide metabolism. No deficiency effects are known in humans. Toxic if inhaled; excess results in neurological disorders
Vanadium (V)[c]	0.02	0.3	1971; believed to be essential for growth
Selenium (Se)	0.02	0.3	1957; part of glutathione peroxidase. Deficiency can lead to cardiomyopathy (Keshan disease in China)
Barium (Ba)[b]	0.02	0.3	
Arsenic (As)[c]	0.01	0.1	1975; believed to be essential to growth
Boron (B)[c]	0.01	0.1	
Nickel (Ni)[c]	0.01	0.1	1971; interferes with iron absorption. Excess exposure causes eczema and cancer
Chromium (Cr)	0.005	0.7	1959; believed to activate insulin. Occupational hazards can lead to allergy, eczema and cancer
Cobalt (Co)	0.003	0.04	1935; component of vitamin B_{12}
Molybdenum (Mo)	<0.005	<0.07	1953; linked to xanthines, aldehyde and sulfide oxidases. Deficiency symptoms not known in humans
Lithium (Li)[c]	0.002	0.03	

[a] Adapted from W. Kaim and B. Schwederski, *Bioinorganic Chemistry: Inorganic Elements in the Chemistry of Life. An Introduction and Guide*, Wiley, Chichester, 1994, p. 7, and references contained therein

[b] Not essential

[c] Essentiality uncertain

∏   What elements are you aware of in your everyday life?

Examples of these might include lead in petrol, aluminium in drinking water, tungsten in electric light bulbs, etc.

By way of an illustration, the human body is composed of a large number of elements (Table 1.1). Of these elements it is possible to observe that a large number are essential to our existence, i.e. their removal would cause severe, irreversible damage. However, even an essential element can cause problems if its concentration is too high; this can be represented in a general dose–response diagram (Figure 1.1). Thus our physiological state depends upon the concentration of a particular element (or more correctly, the chemical form of the element — chemical speciation) in our food supply. Too low an intake of an essential element and our bodies show signs of deficiency (e.g. iron deficiency leads to anaemia, a lack of fluorine can cause dental caries, and selenium deficiency leads to muscular weakness), while too high a concentration and we will experience toxic effects (e.g. mercury, arsenic and lead poisoning). At either end of these two extremes, death will result. It is essential, therefore, that we maintain the correct intake of each essential element to maintain a proper state of health. In order to keep these balances we are advised by government agencies to ingest recommended dietary allowances of essential elements. Atomic spectroscopy is therefore used to monitor trace elements in food and our water supply.

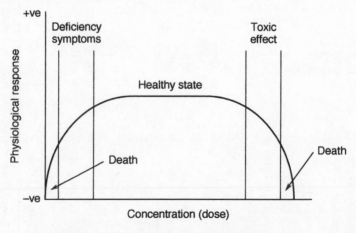

Figure 1.1   Dose–response diagram

1.2   PRECISION AND ACCURACY

These are probably the most often quoted terms in analytical chemistry. In contrast to their common English usage, each has its own strict scientific definition. Precision is defined as the closeness of the grouping of individual results, whereas accuracy is the closeness of the results to the true value. Obviously, the true value may not be known. Both terms can be illustrated by considering the bull's-eyes shown in Figure 1.2a (here we assume that the black dots represent specific results and the aim is to be as close to the centre of the bull's-eye as possible).

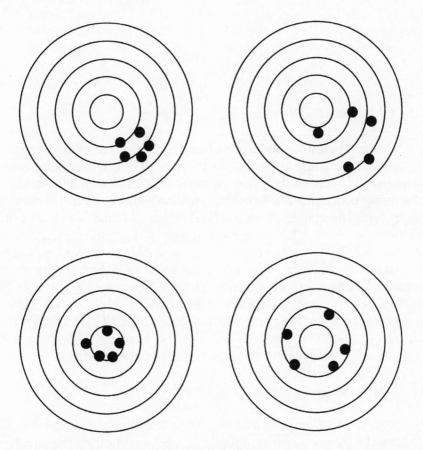

Figure 1.2a   Precision and accuracy; the centres of the bull's-eyes represent the 'true' value

SAQ 1.2

Which of the sets of results shown in Figure 1.2a would you describe as accurate and which as precise? It is possible to come up with suitable answers (i.e. high accuracy, low accuracy, high precision, or low precision) for each bull's-eye.

1.3 CALIBRATION

Calibration requires the establishment of a relationship between signal response and a known set of standards. The standards in atomic spectroscopy refer to the production of a series of aqueous solutions of varying concentration (working standard solutions) of the analyte of interest. The choice of the original standard (from which all other dilutions will be made) is important, as any impurity present in this solution will be transferred to all of the working standard solutions. The standard material may take the form of a high-purity metal (available as powders, wires, rods, etc.) or high-purity-grade salts (or oxides). Commercial suppliers can provide solutions of known concentration for subsequent dilution. In addition, it is important that accurate transfer of the standard solution is made. This should be carried out by using calibrated pipettes and volumetric flasks.

∏ How would you calibrate a pipette?

The weight of 1.00 ml of water is 1.00 g; weighing the actual volume dispensed is therefore an acceptable means of calibrating the pipette.

This also has consequences for solid samples which require dissolving and making up to known volumes by using volumetric flasks. All standards should be stored under appropriate conditions. This means the use of pre-acid-cleaned glass flasks at 4°C.

The mathematical relationship most commonly used for calibration is of the following form:

$$y = mx + c \qquad (1.1)$$

where y is the signal response, i.e. the absorbance in the case of atomic absorption spectroscopy, x is the concentration of the working solution (in appropriate units, e.g. $\mu g\,ml^{-1}$ or parts per million (ppm)), m is the slope of the graph, and c is the intercept on the x-axis.

By measuring the signals for a series of working solutions of known concentrations it is possible to construct a suitable graph (Figure 1.3a). Then, by presenting a solution of unknown concentration to the instrument, a signal is obtained which can be interpreted from the graph, thereby determining the concentrations of the element in the unknown solution.

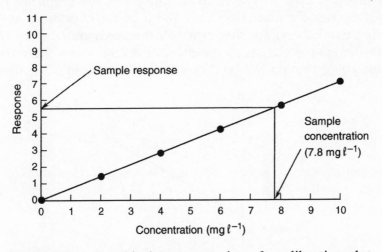

Figure 1.3a   Graphical representation of a calibration plot

Straight lines are not always obtained. If the scatter on the points on the graph is erroneous this may be due to either an intermittent instrument fault, or perhaps, which is more likely, to poor technique in preparing the working standards solutions. It should be noted, however, that calibration plots, particularly in atomic absorption spectroscopy, are frequently not linear, and are more likely to be curved. The fitting of data to these curves requires the use of linear regression analysis, which is beyond the scope of this present text.

An alternative approach to performing a direct calibration (as has just been described) is the use of the method of standard additions. This may be particularly useful if the sample is known to contain a significant portion of a potentially interfering matrix. In standard additions, the calibration plots no longer pass through zero (on both the *x*- and *y*-axes). As the concept of standard additions is to eliminate any matrix effects present in the sample, it is not surprising to find that the working standard solutions all now contain the same volume of the sample, i.e. this same sample volume is introduced into a succession of working solutions. Each of the working solutions, containing the same volume of the sample, is then introduced into the instrument and the response is once again recorded as before. However, plotting the signal response (e.g. the absorbance in atomic absorption spectroscopy) against the analyte concentration produces a very different type of graph. In this situation, the graph no longer passes through zero on either axis, but if correctly drawn, the graph can be extended (extrapolated) towards the *x*-axis until it intercepts it. By maintaining a constant concentration on the *x*-axis the unknown sample concentration can be determined (Figure 1.3b). It is essential

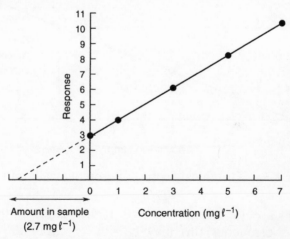

Figure 1.3b   Graphical representation of a standard additions plot

that this graph is linear over its entire length, otherwise considerable error can be introduced.

A number of self-assessment questions are given below to allow both types of graph to be plotted and results obtained. You should note that dilution factors need to be determined and used in order to obtain the correct values. (It may be advantageous to consider these questions after you have completed the chapters relating to the techniques mentioned.)

The calculation of a dilution factor can be illustrated as follows. An accurately weighed (0.5235 g) soil sample is digested in 25 ml of concentrated nitric acid, cooled and then quantitatively transferred to a 100 ml volumetric flask and made up to the mark with high-purity water. This solution is then further diluted by taking 10 ml of the solution and transferring it to another 100 ml volumetric flask, where it is again made up to the mark with high-purity water. The dilution factor is given by the following:

$$\frac{100\,\text{ml}}{0.5230\,\text{g}} \times \frac{100\,\text{ml}}{10\,\text{ml}} = 1912\,\text{ml}\,\text{g}^{-1} \qquad (1.2)$$

If the solution was then analysed and found to be within the linear portion of the graph, the value for the dilution factor would then be multiplied by the concentration obtained from the graph, thus producing a final value which is representative of the element under investigation.

∏ If the concentration of an element from the plotted graph is found to be 15 μg ml^{-1}, then what is the concentration of the element in the original soil sample?

This is obtained by multiplying the dilution factor by the element concentration, i.e.

$$\frac{1912\,\text{ml}}{\text{g}} \times \frac{15\,\mu\text{g}}{\text{ml}} = 28\,680\,\mu\text{g}\,\text{g}^{-1} \qquad (1.3)$$

This result can also be expressed in terms of other units, e.g. 28 680 mg kg^{-1} or 2.87 wt%.

SAQ 1.3

(a) A River Tyne water sample was analysed by inductively coupled plasma (ICP) atomic emission spectroscopy (AES) for Ni. The sample was nebulised directly and the signal obtained was 15.6 mV. A calibration plot was generated by using 0, 1, 5, 10, and 50 ppm solutions, which gave the following responses, i.e. 0, 2, 9, 18, and 93 mV. What is the concentration of Ni in the original River Tyne water sample?

(b) A coastal sea water sample was analysed by ICP-AES for Cu. The sample was diluted by placing 10 ml in a 100 ml volumetric flask and adding 90 ml of distilied water. A calibration plot was generated by diluting a 1000 ppm stock solution. 10 ml of the stock solution were then placed in a 100 ml volumetric flask and made up to the mark with distilled water (working solution). This solution was diluted in series as follows:

Flask	Cu working solution (ml)	Water (ml)	Total volume (ml)
1	0	100	100
2	1	99	100
3	2	98	100
4	3	97	100
5	5	95	100

SAQ 1.3
(Contd)

The signals obtained were as follows:

Flask	Signal (mV)
1	0
2	150
3	290
4	435
5	730
Diluted sample	490

What is the concentration of Cu in the original coastal sea water sample?

(c) 0.5020 g of a steel sample was digested in concentrated acid and then transferred to a volumetric flask (100 ml) and made up to the mark with distilled water. The sample was then diluted 10 times. The diluted sample was then analysed for Pb as follows:

Flask	Volume of 100 ppm Pb solution (ml)	Digested and diluted sample (ml)	Volume of water (ml)	Total volume (ml)
1	0	20	80	100
2	1	20	79	100
3	2	20	78	100
4	3	20	77	100
5	5	20	75	100
6	7	20	73	100

→

SAQ 1.3
(Contd)

After analysis, the following results were obtained:

Flask	Signal (mV)
1	29
2	37
3	44
4	52
5	68
6	83

Calculate the concentration of Pb in the original sample in units of $\mu g\,g^{-1}$ and wt%.

(d) A sample of soil was accurately weighed (0.5250 g) into a microwave vessel and 9 ml of concentrated HNO_3 and 3 ml of concentrated HF were added. After heating for 20 min in a microwave oven the sample was allowed to cool. The contents of the vessel were quantitatively transferred to a 100 ml volumetric flask and analysed for Cu and Ni by using an ICP-AES instrument. Calibration of the instrument was carried out by using suitable dilutions of 1000 $\mu g\,ml^{-1}$ Cu and Ni stock solutions.

Initially, 10 ml of each stock solution were diluted by addition to a 100 ml volumetric flask and the appropriate amount of dilute acid added (diluted stock solution). Then, the (diluted stock) solutions were diluted further, according to the following procedure, to obtain a series of calibration solutions:

SAQ 1.3
(Contd)

Flask	Diluted stock solution/ volume added to a 100 ml volumetric flask (ml)
1	0
2	1
3	5
4	10
5	20

Using the calibration solutions, the digested sample was analysed, with the following results being obtained:

Flask	Cu ICP-AES signal (mV)	Ni ICP-AES signal (mV)
1	0	0
2	7	15
3	28	45
4	53	104
5	101	205
Digested sample solution	82	75

Calculate the concentrations (in $mg\,kg^{-1}$) of nickel and copper in the original soil sample.

(e) When a sample of River Tyne water was analysed for Pb by direct aspiration into an atomic absorption spectrometer equipped with an air–acetylene flame an absorbance reading of 0.25 was obtained. A calibration plot was prepared by diluting a standard $1000\,\mu g\,ml^{-1}$

→

SAQ 1.3
(Contd)

solution $10 \, \mu g \, ml^{-1}$ and then taking the following volumes: 0, 10, 20, 30, and 50 ml. Each aliquot was then diluted to 100 ml with water and analysed by flame atomic absorption spectroscopy (FAAS). The following results were obtained:

Diluted stock solution/ $10 \, \mu g \, ml^{-1} \, Pb$	Absorbance reading
0	0.000
10	0.082
20	0.162
30	0.245
50	0.410

Calculate the concentration (in $\mu g \, ml^{-1}$) of Pb in the sample.

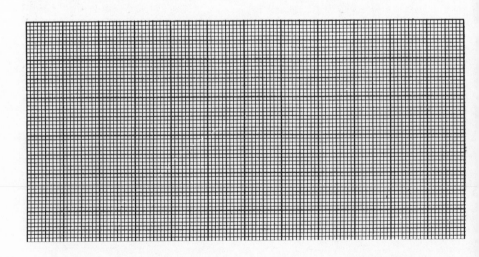

SAQ 1.3

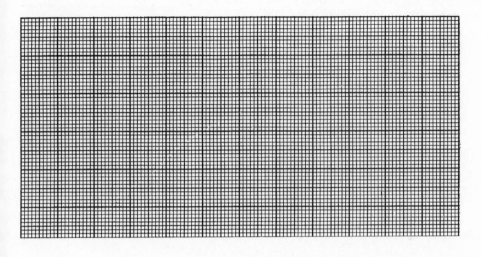

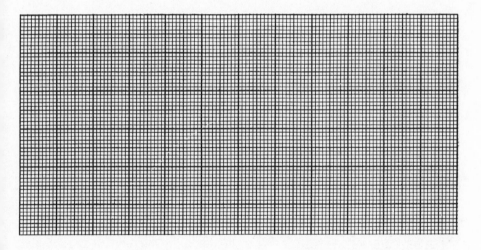

SAQ 1.3

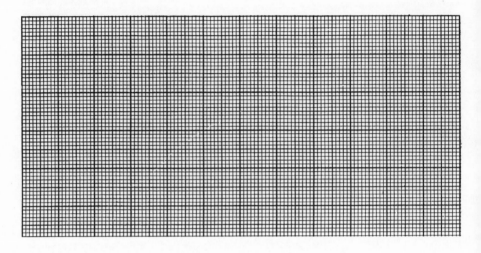

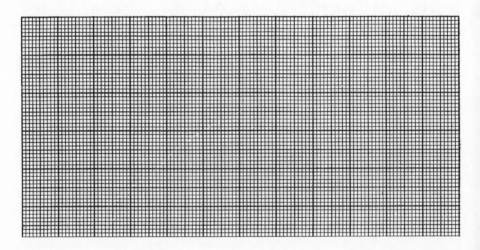

There is a third type of calibration procedure available, which is known as isotope dilution analysis. This technique is only applicable to mass spectrometry; details are given elsewhere in this text (see Section 6.4).

1.4 DETECTION LIMITS

The detection limit is defined as the minimum concentration that can be detected by the analytical method with a given certainty. This is often taken as the mean value of the blank, plus three times its standard deviation. This term is often used to compare one instrumental technique with another. Chapter 7 compares the detection limits of the atomic spectroscopic techniques described in this book. Figure 1.4 shows some examples of signals which are close to the detection limit.

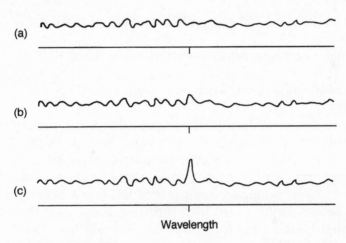

Figure 1.4   Signals close to the detection limit

∏   Which of the responses in Figure 1.4 would you consider to be a definite signal?

The most likely signal occurs in example (c). No signal occurs in (a), with the response only appearing as background noise. In example (b), while it may appear that a small signal occurs at the specified wavelength, this is not, in fact, discernible from the background noise.

1.5 QUALITY CONTROL

It has been said that most of the vast amount of data generated in laboratories is probably worthless! It is probable that this data is worthless, not because of the fact that the instrument was not functioning properly, but because no established protocol was prepared in advance of the analytical procedure. It is therefore essential that a good quality assurance scheme is operated in all laboratories. The main objectives of a quality assurance programme should be as follows:

● to maintain a high quality of laboratory performance;

● to ensure the quality of data produced;

● to identify good and appropriate methods of analysis;

● to maintain and upgrade analytical instruments;

● to ensure good record keeping of methods and results.

The maintenance of data quality can be achieved, in part, by including (certified) reference materials in the analytical protocol. A certified reference material is a substance for which one or more elements (in this case) have known values, with estimates of their uncertainties, produced by a technically valid procedure; this is accompanied by a traceable certificate, issued by a certifying body. Typical examples of certifying bodies are the National Institute for Standards and Technology (NIST), based in Washington DC, USA, the Community Bureau of Reference (BCR), Brussels, Belgium, and the Laboratory of the Government Chemist (LGC), London, UK. The accompanying certificate, in addition to providing details of the certified elemental concentrations and their uncertainties, also provides other details of the sample, minimum sample weights to be used, storage conditions, moisture content, etc. An example of such a certificate is shown in Figure 1.5.

Summary

Analytical measurements have many features that need to be considered in order to derive meaning from the results. The terms accuracy and precision are frequently misused in the English language. However, in scientific language their definition is unequivocal. The

National Institute of Science and Technology

Certificate of Analysis

Standard Reference Material 1572

Citrus Leaves

Certified Values of Constituent Elements[1]

Major and Minor Constituents

Element	Content (wt%)[2]
Calcium	3.15 ± 0.10
Magnesium	0.58 ± 0.03
Phosphorus	0.13 ± 0.02

Trace Constituents

Element	Content (μg g^{-1})[2]
Aluminium	92 ± 15
Arsenic	3.1 ± 0.3
Cadmium	0.03 ± 0.01
Lead	13.3 ± 2.4
Nickel	0.6 ± 0.3
Zinc	29 ± 2

[1]The certified values for the constituent elements are based on results obtained by either definitive methods of known accuracy or by two or more independent analytical methods.

[2]The values are based on dry weight. Samples of this Standard Reference Material (SRM) must be dried before weighing and analysis by either of the following procedures: (a) drying for 2 h in air in an oven at 85°C; (b) drying for 24 h at 20–25°C.

Figure 1.5 An example of a certificate of analysis for elements in citrus leaves

acquisition of meaningful data relies on the correct application of the selected calibration strategy. This, in turn, relies upon an understanding of the operating principles of the analytical instrument in question and the limit of its sensitivity. In order to further promote the validity of the analytical measurements, adherence to a quality assurance protocol is necessary.

Objectives

On completion of this chapter you should be able to:

- understand the importance of trace element analysis;

- identify accuracy and precision;

- plot an appropriate graph and interpret it;

- calculate and evaluate dilution factors;

- determine concentrations for a range of units;

- appreciate the significance of the detection limit;

- appreciate the concept of quality control and the use of reference materials.

2. Sample Preparation

Sample preparation is probably the single most neglected area in analytical chemistry. While the level of sophistication of the instrumentation for analysis has increased significantly, a comparatively low technical basis of sample preparation often remains. It is normal in atomic spectroscopy for the sample to be found in one of two forms, i.e. solid or liquid. In this present case, the latter would seem to be the easiest form in which to handle the sample, with maybe a requirement for filtration being all that is required. However, the inherent lack of sensitivity of many atomic spectroscopic techniques and the need (environmental protection, compound impurities, etc.) to carry out determinations at lower and lower levels means that invariably some form of preconcentration is required.

∏ Consider how you might preconcentrate a liquid sample, e.g. an aqueous solution.

The easiest way to evaporate or remove the water by heating.

If, on the other hand, the sample is in a solid form, the normal requirement is to convert the solid material (biological, environmental or geological) into the liquid form. As you will see later (Section 5.3.3), it is possible to analyse solids directly by using atomic spectroscopy, but this is not the preferred approach. (Note: X-ray fluorescence spectroscopy is an alternative approach for the analysis of solid samples, but this is not the subject of this present text.) Conversion of a solid matrix into a liquid matrix involves the decomposition of the sample.

∏ How might you decompose or dissolve a solid sample?

You could use heat or acid, or both.

The major problem in preparing samples for trace element analysis is the risk of contamination. The latter is a particular problem, with several probable causes, i.e. the grade of reagents used, the vessels used for storage, digestion or dilution of the samples and their previous history, and human intervention.

∏ Which element is a major component of talcum powder?

Aluminium — so if you have liberally dosed your body in talc that morning there is a good chance that some might drop into the vessel containing the sample to be analysed.

∏ How might you prevent accidental contamination of a sample?

Wearing a fastened laboratory coat would be a good first step towards this.

It is imperative for trace analytical work that the grade of reagents used is of the highest purity (this also includes the water supply used for sample dilution). Major suppliers of reagents have classifications that allow the purchase (at a higher cost) of purer reagents. The elimination of impurities from solid reagents is particularly troublesome, and so the higher-cost reagent will not be totally pure. However, what it does contain in the way of impurities will have already been characterised and the details quoted with the suppliers' information. The use of sample blanks in the analytical procedure is essential for identifying any problem elements. The nature of the vessels used for storage, dissolution or subsequent dilution and their previous history is also important. The utilisation of any type of container should raise immediate concern as to its cleanliness and the potential leaching out of any trace elements. It is common practice to soak sample vessels in a suitable acid leaching bath (e.g. 10% nitric acid) for at least 24 h, followed by rinsing with copious amounts of ultrapure water.

∏ What type of water would you consider to be ultrapure?

Drinking water, commercially bottled drinking water or laboratory distilled water might all be regarded as possibilities. The actual answer depends to some extent on the work to be carried out. For ultratrace analysis, it is possible to purchase commercial systems that filter (distilled) water through a combination of ion-exchange columns to remove trace element impurities.

Concern should then be focused upon the method of storage of the vessel prior to its use. There is no use preparing an open vessel in this time-consuming manner and then allowing it to be stored on a shelf where dust can accumulate inside the container. Dust is a major source of trace metals, e.g. Aluminium. This point leads directly into the question of the human factor. 'We' are a major source of contamination in trace element analysis. Therefore, as well as the normal laboratory safety associated with wearing a laboratory coat and safety glasses, it may be necessary to take additional steps. It would be difficult to identify a person who has never suffered from dandruff (a major source of trace metals) or sweaty hands (salts), to name just two of the obvious outpourings from the body. Thus, for trace element work it may be necessary to take additional precautions and wear 'contaminant'-free gloves and a close fitting hat. These precautions may sound a bit 'over-the-top' for the normal everyday laboratory, but consider the precautions that are required today in the semiconductor/nuclear industry.

2.1 TRACE ELEMENT PRECONCENTRATION

Quite often the analyte (element) of interest is present in the sample at a level which is below or close to the limit of detection of the system being used. In this situation, unless you can introduce the sample into another instrument which is more sensitive for the analyte of interest, the requirement is to preconcentrate the analyte. This situation is most likely to occur for aqueous-based samples. In the case of a solid sample it may be possible to digest more sample, thereby presenting more of the analyte to the instrument. Various approaches to trace element preconcentration are available and these are discussed below.

As well as preconcentration it may be necessary to carry out a separation process to remove an interference or to distinguish between different chemical forms of an element, e.g. its oxidation state. Both of these approaches are commonly found in the published scientific literature.

2.1.1 Chromatographic

Of all the different types of chromatographic separation techniques that are available, the most common type for preconcentration is ion-exchange chromatography. In this present case, cation- or anion-exchange chromatography are both commonly used. As their names suggest, cation exchange is used to separate metal ions (positively charged species), while anion exchange is used to separate negatively charged species. At first sight, it may seem that the only useful form of ion-exchange chromatography is cation exchange, but this is not always the case.

∏ Can you suggest cases where anion-exchange chromatography may be used?

Some elements are found as anions, e.g. SO_4^{2-} for sulfur determination, and arsenite (AsO_2^-) and arsenate (AsO_4^{3-}) for arsenic determination.

As an example of the use of a strong cation-exchange resin, the following general equations can be written.

1. Metal ion (M^{n+}) preconcentrated on cation-exchange resin:

$$nRSO_3^-H^+ + M^{n+} = (RSO_3^-)nM^{n+} + H^+ \qquad (2.1)$$

2. Desorption of metal ion using acid:

$$(RSO_3^-)nM^{n+} + H^+ = nRSO_3^-H^+ + M^{n+} \qquad (2.2)$$

In this way, a selected metal ion can be isolated (separated) and preconcentrated from its matrix. This process can be carried out in two ways, namely batch, or column. In the batch process the ion-exchange resin is added to the aqueous sample, whereas in the other process the resin is packed into a chromatographic column. The former would be always carried out off-line, whereas the latter could be carried out in either the on-line or off-line mode.

2.1.2 Solvent Extraction

Solvent extraction, or liquid–liquid extraction, is historically the most often used method of preconcentration. This involves two immiscible liquids, e.g. water and an organic solvent, which are shaken together so that the metal of interest in the larger-volume water sample is relocated into the smaller volume of the organic phase. In order to persuade the metal to transfer from the aqueous phase to the organic phase the use of a chelate or ligand is often required. The most commonly used chelate in atomic spectroscopy is ammonium pyrrolidine dithiocarbamate (APDC), with methyl isobutyl ketone (MIBK) as the organic solvent. As APDC forms stable metal complexes over a wide range of pH (Table 2.1a), its universal acceptance is assured.

Table 2.1a pH dependence of APDC complexes[a]

pH Range	Elements that form APDC complexes
2	W
2–4	Nb, U
2–6	As, Cr, Mo, V, Te
2–8	Sn
2–9	Sb, Se
2–14	Ag, Au, Bi, Cd, Co, Cu, Fe, Hg, Ir, Mn, Ni, Os, Pb, Pd, Pt, Ru, Rh, Tl, Zn

[a]Data taken from G.F. Kirkbright and M. Sargent, *Atomic Absorption and Fluorescence Spectroscopy*, Academic Press, London, 1974

SAQ 2.1	Is it possible to separate tin (Sn) from lead (Pb) by forming an APDC complex?

Other common chelating agents which are used in solvent extraction are shown in Table 2.1b.

Table 2.1b   Common chelating agents used for solvent extraction

Chelating agent	Structure
Ammonium pyrrolidine dithiocarbamate (APDC)	
Sodium diethyldithiocarbamate (NaDDC)	
8-Hydroxyquinoline (oxine)	
2,4-Pentanedione or acetylacetone (acac)	

2.1.3   Other Methods

While other techniques, such as co-precipitation, can be used for preconcentration they are not as common as those discussed above. In fact, if it was possible to determine an element at the required level (i.e. to national or international guidelines) in the matrix, the analyst would prefer not to carry out any preconcentration. By the very nature of the preconcentration process, it is both a labour-intensive and time-consuming operation. In addition, it may preconcentrate potential interferences as well as the analyte(s) of interest.

2.2   DECOMPOSITION TECHNIQUES

Decomposition involves the liberation of the analyte (metal) of interest from an interfering matrix by using a reagent (e.g. mineral/oxidising acids or fusion flux) and/or heat. The utilisation of reagents (acids) and external heat sources can in itself cause problems. In elemental analysis these problems are particularly focused on the risk of contamination and loss of analytes. It should be borne in mind that complete digestion may not always be required as atomic spectroscopy frequently uses a hot source, e.g. flame or inductively coupled plasma, which provides a secondary method of sample destruction. Therefore, methods that allow sample dissolution may be equally as useful.

∏ Consider the difference between the terms digestion and dissolution.

Digestion infers the complete destruction of the sample matrix, whereas dissolution considers the liberation from the matrix of the analyte of interest (without the requirement for complete destruction of the matrix). In the latter case, it may still be possible to identify 'organic' parts of the matrix by using appropriate techniques.

2.2.1 Acid Digestion (including Microwave)

Acid digestion involves the use of mineral or oxidising acids and an external heat source to decompose the sample matrix. The choice of an individual acid or combinations of acids is dependent upon the nature of the matrix to be decomposed. The most obvious example of this relates to the digestion of a matrix containing silica (SiO_2), e.g. a geological sample. In this situation, the only appropriate acid to digest the silica is hydrofluoric acid (HF). No other acid or combination of acids will liberate the metal of interest from the silica matrix.

SAQ 2.2	Why should HF be so effective for the digestion of silica?

A summary of the most common types of acids used for digestion and their various applications is presented in Table 2.2.

Once the choice of acid is made, the sample is placed in an appropriate vessel for the decomposition stage. The choice of vessel, however, depends upon the nature of the heat source to be applied. Most commonly, the acid digestion of solid matrices has been carried out in open glass vessels (beakers or boiling tubes), using a hotplate or multiple-sample digestor. The latter allows a number of boiling tubes (e.g. 6, 12 or 24 tubes) to be located in the well of a commercial digestor. In this manner, multiple samples can be simultaneously digested. An alternative approach involves the use of microwave heating. In this latter case, the sample is placed inside a (normally) closed vessel made of PTFE (polytetrafluoroethylene) or Teflon PFA (perfluoroalkoxy vinylether). Open microwave systems are also available. A typical microwave system operates at 2.45 GHz with up to 12 sample vessels arranged on a rotating carousel. Commercial systems may have additional features, such as a PTFE-lined cavity, a safety vent (if the pressure inside a vessel is excessive the vent will open, thus allowing the contents to go to waste), and the ability to measure both the temperature and pressure inside a single vessel.

∏ What advantage in terms of digestion time could the use of a microwave system offer over a conventional hot plate?

Digestion is faster when using a microwave oven (consider the domestic microwave oven and a conventional electric or gas cooker).

The use of open vessels for digestion can lead to additional problems associated with loss by volatilisation of the element species. This can be rectified by the correct choice of both the reagents and the type of digestion apparatus to be used.

2.2.2 Dry Ashing

Probably the simplest of all decomposition systems involves the heating of the sample in a muffle furnace in the presence of air at 400–800°C. After decomposition, the residue is dissolved in acid and transferred to a volumetric flask prior to analysis. This allows organic matter to be destroyed. However, the method may also lead to the loss of volatile

Table 2.2 Common acids and mixtures used for decomposition

Acid(s)	Boiling point (°C)	Comments
Hydrochloric (HCl)	110	Useful for salts of carbonates, phosphates, some oxides, and some sulfides. A weak reducing agent; not generally used to dissolve organic matter
Sulfuric (H$_2$SO$_4$)	338	Useful for releasing a volatile product; good oxidising properties for ores, metals, alloys, oxides and hydroxides; often used in combination with HNO$_3$. **Caution**: H$_2$SO$_4$ must never be used in PTFE vessels (PTFE has a melting point of 327°C and deforms at 260°C)
Nitric (HNO$_3$)	122	Oxidising attack on many samples not dissolved by HCl; liberates trace elements as the soluble nitrate salt. Useful for dissolution of metals, alloys and biological materials
Perchloric (HClO$_4$)	203	At fuming temperatures, a strong oxidising agent for organic matter. **Caution**: violent, explosive reactions may occur — care is needed. Samples are normally pre-treated with HNO$_3$ prior to addition of HClO$_4$
Hydrofluoric (HF)	112	For digestion of silica-based materials; forms SiF$_6^{2-}$ in acid solution; caution is required in its use; glass containers should not be used, only plastic vessels. Protective clothing/eyewear essential; in case of spillages, calcium gluconate gel (for treatment of skin contact sites) should be available prior to usage; evacuate to hospital immediately if skin is exposed to liquid HF
Nitric/hydrochloric (HNO$_3$/HCl)	–	A 1/3 mixture of HNO$_3$/HCl is called aqua regia; forms a reactive intermediate, NOCl. Used for metals, alloys, sulfides and other ores; best known because of its ability to dissolve Au, Pd and Pt

elements, e.g. Hg, Pb, Cd, Ca, As, Sb, Cr, and Cu. Therefore, while compounds can be added to retard the loss of volatiles, its use is limited.

2.2.3 Fusion

Some substances, e.g. silicates and oxides, are not normally destroyed by the action of acid. In this situation, an alternative approach is required. Fusion involves the addition of an excess (10-fold) of reagent to the (finely ground) sample, which is placed in a metal (e.g. Pt) crucible, followed by heating in a muffle furnace (300–1000°C). After heating for a period of time (from a few minutes to several hours) a clear 'melt' should result, thus indicating the completion of the decomposition process. After cooling, the melt will then dissolve in a mineral acid. Typical reagents include sodium carbonate (heat to 800°C; dissolve with HCl), lithium meta- or tetraborate (heat to 900–1000°C; dissolve with HF), and potassium pyrosulfate (heat to 900°C; dissolve with H_2SO_4). The obvious addition of excess reagent (flux) can lead to a high risk of contamination. In addition, the high salt content of the final solution may lead to nebuliser problems, i.e. blockages (see Section 5.3.1) in the subsequent analysis.

Summary

Sample preparation makes probably a greater contribution to poor analytical data than any other procedure. The variability that can be incurred may be attributed to many causes in the preconcentration and/or decomposition of the samples. Therefore, it is essential to select the appropriate sample preparation method for both the sample and the analytical measurement technique being employed. This should not be a 'marriage of convenience', but be designed to obtain the maximum possible chemical information from the process.

Objectives

On completion of this chapter you should be able to:

● identify appropriate sample preparation techniques for various sample types;

- identify the steps required in the laboratory to prepare representative samples for analysis;

- identify sources of contamination;

- assess the different types of trace element preconcentration techniques;

- assess the different types of decomposition methods.

3. The Theory of Atomic Spectroscopy

3.1 HISTORICAL PERSPECTIVE

The term spectroscopy was originally used in the context of visible light and its subsequent resolving into component wavelengths to produce spectra. The earliest forms of this were investigated in 1666 by Isaac Newton who discovered that sunlight could be divided into spectral colours, i.e. the colours of the rainbow (red, orange, yellow, green, blue, indigo and violet). In 1802, Wollaston, while observing the sun's solar spectrum, noticed that dark lines were present; in other words, the spectrum was not continuous. It was left to Fraunhofer (1814) to further investigate these dark lines in the sun's spectrum; he was subsequently able to observe over 700 of them. These lines have come to be known as Fraunhofer lines, and result from the absorption of radiation by atoms.

In 1826, Talbot and Herschel introduced metal salts of sodium, potassium, lithium and strontium into a flame and observed flame colours through a spectroscope. The colours observed in the flame occurred at the same wavelengths as those seen by Fraunhofer. It was apparent that these lines were emission lines. The suggestion was made that the emission lines observed in the flame could be used as an appropriate method for identification. It was left to Kirchoff and Bunsen (1860) to show that such emission lines were due to atoms and not compounds. With the advent of the pre-mixed air–coal gas flame (or Bunsen burner) the technique of spectrochemical analysis was introduced for the identification of elements.

3.2 INTRODUCTION

Subsequent developments have expanded the use of the term spectroscopy to include not only visible radiation but other parts of the electromagnetic spectrum, such as X-rays, ultraviolet, infrared, microwaves and radiowaves. All spectroscopic techniques depend upon the emission or absorption of electromagnetic radiation, which arises from certain energy changes within a molecular, or, in the case of this present text, an atomic system. The regions of the electromagnetic spectrum can be identified in terms of a wavelength and a frequency (Figure 3.2). (Note: the wave–particle duality of light is not discussed in this book, but most, if not all, textbooks on analytical chemistry contain the appropriate information.) A mathematical relationship

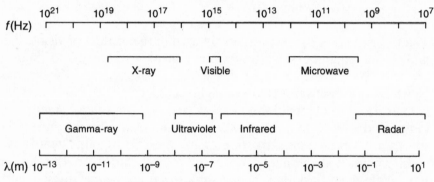

Figure 3.2   Regions of the electromagnetic spectrum

exists that allows wavelength (λ) and frequency (f) to be determined, provided one of the terms is known. Wavelength is normally expressed in terms of metres (m) and frequency in cycles per second (s^{-1}) or hertz (Hz). The relationship, which also incorporates a constant, i.e. the velocity of light (c), is as follows:

$$c = f \times \lambda \qquad (3.1)$$

where c approximates to 3.00×10^8 ms^{-1}.

SAQ 3.2a	If the frequency of electromagnetic radiation is 5×10^{14} Hz, what is the wavelength of this radiation and in which spectral region does is occur?

SAQ 3.2a

It should be noted here that it is common practice to use prefixes or notations to remove powers from numerical values, e.g. millimetre (mm) is the same as 0.001 m or 10^{-3} m. Likewise, numerical prefixes are used to represent both negative and positive powers (Table 3.2).

Table 3.2 Common prefixes and their representations

Number	Power representation	Prefix and name
1000 000 000	10^9	giga, G
1000 000	10^6	mega, M
1000	10^3	kilo, k
0	–	–
0.001	10^{-3}	milli, m
0.000 001	10^{-6}	micro, μ
0.000 000 001	10^{-9}	nano, n
0.000 000 000 001	10^{-12}	pico, p

As well as frequency and wavelength, electromagnetic radiation can also be expressed in terms of packets of energy (E) called photons (or quanta). The energy of a photon (in units of joules or J) can be expressed, again by a mathematical relationship, in terms of frequency. In this relationship, a further constant, known as the Planck constant (h, 6.626×10^{-34} Js) is used, so that:

$$E = hf \tag{3.2}$$

By the substitution of equation (3.1) we can also have an expression which is related directly to wavelength:

$$E = hc/\lambda \tag{3.3}$$

SAQ 3.2b   Calculate the energy in joules of one photon of the radiation derived in SAQ 3.2a.

3.3 ORIGINS OF SPECTRAL TRANSITIONS

If an atom is supplied with sufficient energy (thermal or electrical), the electron is raised from a low energy level (e.g. the ground state) to one with a higher energy (an excited state). This is referred to as absorption. As the excited state is unstable, the electron will return to a lower energy state (by inference, a more stable situation). This is referred to as emission. Both absorption and emission occur at certain selected wavelengths, frequencies or energies (Figure 3.3).

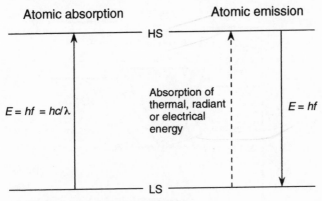

LS = lower energy state or ground state
HS = higher energy state

Figure 3.3 Schematic representation of atomic absorption and atomic emission energy transitions

3.4 ATOMIC SPECTRA

At room temperature, all the atoms of a sample are in the ground state. For example, the single outer electron of sodium occupies the 3s orbital (note that the electron configuration for Na is $1s^2\ 2s^2\ 2p^6\ 3s^1$). In a hot environment, sodium atoms are capable of *absorbing* radiation, such that electronic transitions from the 3s level to higher excited states can occur. These electronic transitions occur at specific wavelengths. Experimental observation of sodium identifies absorption peaks at 589.0, 589.6, 330.2, and 330.3 nm. By considering the energy-level diagram shown in Figure 3.4a for sodium, it is possible to identify that these wavelength doublets correspond to electronic transitions from the 3s level to either the 3p or 4p levels for 589.0/589.6 nm and 330.2/330.3 nm, respectively. While other electronic transitions are possible, the strongest, i.e. the most intense, absorption spectrum occurs for electronic transitions from the ground state (3s) to upper levels. The wavelengths at which these transitions occur are called resonance lines.

Similarly, in the hot environment of a flame or plasma, the electron is easily excited to an upper energy level. However, as the lifetime of the excited atom is brief (typically 10^{-8} s) its return to the ground state is

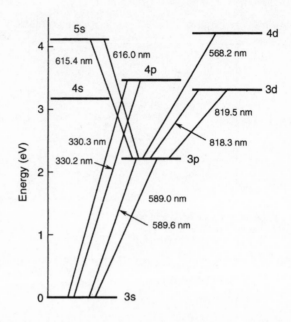

Figure 3.4a   Energy-level diagram for sodium

accompanied by the emission of a photon of radiation. In Figure 3.4b, the wavelength doublet at 589.0 and 589.6 nm represents the most intense emission lines for sodium and these are responsible for the yellow colour observed when sodium salts are introduced into a flame, e.g. a bunsen burner.

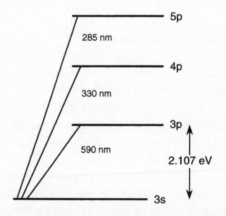

Figure 3.4b   Simplified energy-level diagram for sodium

SAQ 3.4   Confirm that the energy difference between the 3p and 3s levels in Figure 3.4b corresponds to the expected wavelength (note: $1\,eV = 1.602 \times 10^{-19}$ J).

It is important to note that both the emission and absorption lines for sodium occur at identical wavelengths since the transitions involved are between the same energy levels.

Emission spectra, in particular, can be further complicated by the presence of both band and continuous spectra. Band (or molecular) spectra arise from the excitation of molecular species in the hot environment of the flame. Thus, it is common to observe band spectra from diatomic molecules. In order to differentiate these molecular species from any atomic species of interest it is necessary that a high-resolution spectrometer is used (see Section 5.4). Examples of typical molecular species that can be encountered are C_2 molecules (if organic solvents are introduced) and OH radicals. Their appearance can be troublesome, producing undesirable interference effects. Spectra continua which can arise from recombinations, such as molecules in flames, or electrons and ions to form atoms and bremsstrahlung (continuous background emission) in plasmas, generally cause elevation of the background against which the emission lines are measured.

3.5 SPECTRAL LINE INTENSITY

Spectral line intensities depend on the relative populations of the ground or lower electronic state and the upper excited state. The relative populations of the atoms in the ground or excited states can be expressed in terms of the Boltzmann distribution law, as follows:

$$N_1/N_o = g_1/g_o \exp(-\Delta E/kT) \qquad (3.4)$$

where N_1 is the number of atoms in the excited state, N_o is the number of atoms in the ground or lower state, and g_1 and g_o are the number of energy levels having the same energies for the upper (excited) and lower (ground) energy levels, respectively (note: energy levels of the same energy are usually referred to as being degenerate); ΔE is the difference in energy between the lower and upper energy states, k is the Boltzmann constant $(8.314\,\mathrm{J\,K^{-1}\,mol^{-1}})$, and T is the temperature.

Using sodium as an example, there are two degenerate 3p energy levels (excited states) and a single ground state, i.e. 3s, producing a g_1 value of 2 and a g_o value of 1. Simplification of equation (3.4) gives:

$$N_1/N_o = 2 \exp(-\Delta E/kT) \qquad (3.5)$$

At a typical flame temperature of 2500 K, the ratio of the populated upper and ground transition states can be calculated for the spectral transition at 589 nm.

You will recall that $E = hc/\lambda$. By using this equation it is possible to determine the energy of this spectral transition:

$$E = \frac{6.626 \times 10^{-34} \text{ J.s} \times 3.00 \times 10^{8} \text{ m.s}^{-1}}{589 \times 10^{-9} \text{ m}}$$

which gives $E = 3.37 \times 10^{-19}$ J

or for 1 mole of photons:

$$3.37 \times 10^{-19} \text{ J} \times 6.022 \times 10^{23} \text{ mol}^{-1}$$

giving $E = 203\,544$ J mol^{-1}.

By substitution into the revised Boltzmann equation:

$$N_1/N_0 = 2 \exp\left(-203\,544 \text{ J mol}^{-1}/8.314 \text{ J K}^{-1} \text{ mol}^{-1} \times 2500 \text{ K}\right)$$
$$N_1/N_0 = 2 \exp\left(-9.79\right)$$

and therefore:

$$N_1/N_0 = 1.12 \times 10^{-4}.$$

Thus, at a typical flame temperature of 2500 K only 1 atom in 11 200 is in the excited state. (Note that 2500 K is the temperature of the acetylene–air flame commonly used for atomic absorption spectroscopy; see Section 4.4.1.)

SAQ 3.5

A typical temperature of an inductively coupled plasma is 7000 K (see Section 5.2.1). What is the value of the N_1/N_0 population ratio?

3.6 SPECTRAL LINE BROADENING

From the discussion so far you might have the impression that emission and absorption line profiles are very narrow and occur at particular discrete wavelengths. This is in essence true, but due to other processes that occur the observed spectral lines profiles are invariably broadened. Figure 3.6 shows the typical shape of a spectral line. The linewidth is defined as the width at half the peak height ($\Delta\lambda_{1/2}$). Three main factors influence the linewidths, namely natural, Doppler and pressure broadening.

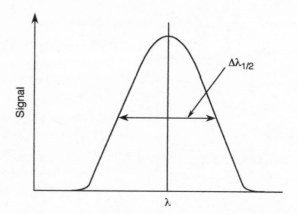

Figure 3.6 Typical shape of a spectral line

The naturalline width broadening is a consequence of the short lifetime (ca. 10^{-8} s) of an atom in an excited state. In 1927, Werner Heisenberg postulated that *nature places limits on the precision with which certain pairs of physical measurements can be made.* Using this *Uncertainty Principle* the natural width of an emission line ($\Delta\lambda_N$) can be determined from the following revised expression:

$$\Delta\lambda_N = \lambda^2 \Delta\nu/c \qquad (3.6)$$

where λ is the wavelength of the emission line, $\Delta\nu$ is the uncertainty in the frequency of the emitted radiation (note that $\Delta\nu$ is equal to $1/\tau$, where τ is the lifetime of the excited state), and c is the speed of light.

SAQ 3.6a For a sodium emission line of wavelength 589 nm and an excited lifetime of 2.5×10^{-9} s, what is the natural linewidth?

It should be noted that the observed linewidth for sodium, at 589 nm, is some 10 times wider than the value found from SAQ 3.6a (assuming that you have carried out the calculation correctly!). Therefore, other broadening processes must predominate. The thermal motion of atoms in a gas (or in a flame or plasma) introduces an additional broadening of the line profile, i.e. Doppler broadening ($\Delta\lambda_{D}$). An equation that describes this broadening is given below (in units of metres):

$$\Delta\lambda_{D} = (2\lambda/c)\sqrt{2RT/M} \tag{3.7}$$

where λ is the wavelength of the emission or absorption, c is the speed of light (3×10^{8} ms^{-1}), R is the gas constant (8.314 JK^{-1} mol^{-1}), T is the temperature, and M is the atomic mass of an atom.

SAQ 3.6b For a sodium atom in a flame at a temperature of 2500 K calculate the Doppler linewidth for the 589 nm spectral line (note that the sodium atomic mass is 23 g mol^{-1}, but in SI units (Système Internationale d'Unités) the value is 23×10^{-3} kg mol^{-1}).

SAQ 3.6b

If you have carried out the above calculation correctly, it should now be evident that the major contributor to the observed linewidth (0.005 nm) for sodium in a flame is Doppler broadening.

Pressure broadening arises from collisions between the emitting or absorbing species with other atoms or ions in the flame or plasma (or hollow-cathode lamp). These collisions cause small changes in the ground-state energy levels and hence a subsequent small variation in the absorbed or emitted wavelength. In flames, the collisions are between the atoms of interest and the combustion products of the flame (note that this type of broadening is usually referred to as Lorentz broadening). This results in a significant broadening which is similar to, but slightly less than that obtained for Doppler broadening (typically 3 pm for sodium).

Summary

The fundamental aspects of atomic spectroscopy are highlighted. Using simple mathematical relationships, the energy, frequency and wavelength can be interconverted to produce numerical values with appropriate units. The basis of atomic spectra is discussed by using sodium as an example. The nature of the spectral line and the processes that lead to spectral line broadening are highlighted.

Objectives

On completion of this chapter you should be able to:

- define the terms absorption and emission of radiation;

- calculate the energy, wavelength or frequency of a spectral line in appropriate units;

- explain what a resonance line is;

- calculate the relative populations of the ground and excited states by using the temperature and the wavelength of the transition;

- define linewidth;

- appreciate the narrowness of atomic spectral lines;

- describe the main types of line broadening in atomic spectroscopy;

- calculate Doppler and natural linewidths.

4. Atomic Absorption Spectroscopy

Atomic absorption spectroscopy (AAS) is probably the most commonly encountered of the techniques described in this book. This is due to the simplicity of the technique and its low capital cost.

4.1 INTRODUCTION

The main components of an atomic absorption spectrometer are a radiation source, an atomisation cell and a method of wavelength selection and detection (Figure 4.1a). The radiation source generates the characteristic narrow-line emission of a selected metal. The atomisation cell is the site where the sample is introduced; the type of atomisation cell can vary but essentially it causes the metal-containing

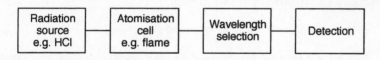

Figure 4.1a   Block diagram of an atomic absorption spectrometer

sample to be dissociated, such that metal atoms are liberated from a hot environment. This environment in the atomisation cell is sufficient to cause a broadening of the absorption line of the metal. The utilisation of the narrowness of the emission line from the radiation source, along with the broad absorption line, means that the wavelength selector only has to isolate the line of interest from other lines emitted by the radiation source (Figure 4.1b). This unique feature of AAS gives it a high degree of selectivity; this process is usually referred to as the 'lock and key' effect.

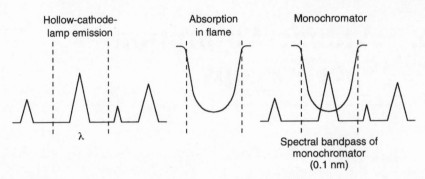

Hollow-cathode-lamp emission   Absorption in flame   Monochromator

Spectral bandpass of
monochromator
(0.1 nm)

Figure 4.1b Principle of atomic absorption spectroscopy (the 'lock and key' effect)

4.2 ABSORPTION OF RADIATION

If a beam of light passes through an atom cell (a specifically defined region that contains atoms), some of the light will be absorbed by those atoms. Figure 4.2 shows the effect on the light beam when it is passed through an atom cell. Several terms can be defined. The *transmittance*, T, of the atom cell is that fraction of the light that is allowed to pass through unaffected; the *absorbance, A*, is then defined as $\log(1/T)$. In addition, the absorption of radiation follows the Beer–Lambert law, so that the absorbance, or *attenuation*, can be expressed by the following relationship:

$$A = -\log T = -\log I/I_0 = abc \qquad (4.1)$$

where I_0 and I are the intensities of the incident and transmitted

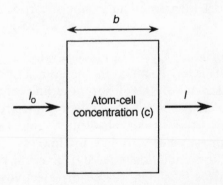

Figure 4.2 Representation of the Beer–Lambert law

light, respectively, *a* is the absorptivity (which is a constant), *b* is the pathlength of the atom cell, and *c* is the concentration of atoms in the cell.

4.3 RADIATION SOURCES

Two types of radiation source have been used for atomic absorption spectroscopy, namely a continuum source or a line source. By far the most common of these is the line source; two types of this source are available, i.e. the hollow-cathode lamp (HCL) and the electrodeless discharge lamp (EDL). The latter is the least favoured line source for AAS. The continuum source will not be discussed in this present book because of its limited applicability.

∏ Is the HCL an emission or absorbance source?

The HCL is an emission source, in that it emits radiation characteristic of a particular element. A hollow-cathode lamp (Figure 4.3), as its name suggests, consists of a hollow cylindrical cathode which is lined with the metal (or metals) of interest. It is possible to purchase multi-element hollow-cathode lamps, but for the purpose of this discussion only single-element lamps will be described. In addition, the HCL contains an anode, which is usually made of tungsten. Both the cathode and anode (the electrodes) are housed in a cylindrical glass envelope which contains a silica window. The HCL is filled with an inert gas, usually argon or neon, under vacuum (100–200 Pa, 1–5 torr). In order

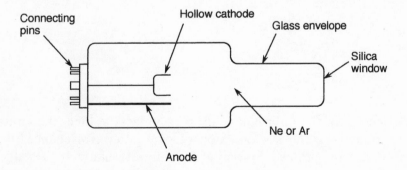

Figure 4.3   Schematic diagram of a hollow-cathode lamp

to initiate the lamp a voltage (100–400 V) is applied across the electrodes, corresponding to a current of 2–30 mA. The passage of the current causes ionisation of the fill gas ($Ar^+ + e^- = Ar^+ + 2e^-$). The positive fill-gas ions are then attracted to the cathode, and upon impact the energy of the incoming fill-gas ions is sufficient to cause metal atoms to be liberated in a process which is known as 'sputtering'.

SAQ 4.3

Write a simple chemical equation to explain the sputtering process that occurs in the hollow-cathode lamp.

The choice of the fill gas depends on two factors. First, the emission spectrum of the fill gas should not coincide with the resonance lines of the metal of interest. Secondly, the fill gas should be capable of causing sputtering of the metal on the cathode. Therefore, the heavier noble gas atom (Ar) is capable of sputtering metals with a higher atomic weight or which have a higher ionisation potential than neon, and vice versa. It is easy to differentiate the fill-gas composition by simple viewing of the HCL in operation. Argon-filled HCLs have a blue discharge, while neon-filled lamps have an orange discharge.

4.4 ATOMISATION CELLS

The atomisation cell is needed to produce ground-state atoms. Several types of atom cell are available, with the most common being the flame cell, although for greater sensitivity a graphite furnace is required. For specialist applications, both hydride generation and cold-vapour techniques are used.

4.4.1 Flames

The types of flames used for AAS can be classified in several ways. However, the commercialisation of this technique has meant that only one type of flame is likely to be encountered, i.e. the pre-mixed laminar flame. In this case, the fuel and oxidant gases are mixed prior to entering the burner (the ignition site) in an expansion chamber.

Various conditions are required for the burner assembly when using a pre-mixed laminar flame, including the following:

- complete mixing of the fuel and oxidant gases should have occurred prior to exiting the expansion chamber;

- the flow of the gases should be laminar as they leave the top of the burner;

- the burner should be able to operate for long periods of time without overheating or clogging;

- the burner should allow a stable flame to be formed with appropriate gas flow rates, thus preventing any flash-back of the flame into the burner/expansion chamber;

- the flame obtained should be an appropriate size and shape to maximise absorption, i.e. have a long atom pathlength (see Section 4.2).

The combustion reactions of the common flames are as follows:

Fuel–Oxidant   Reaction

C_2H_2–air $C_2H_2 + O_2 + 4N_2 \longrightarrow 2CO + H_2 + 4N_2$ (4.2)

C_2H_2–N_2O $C_2H_2 + 5N_2O \longrightarrow 2CO_2 + H_2O + 5N_2$ (4.3)

Within the pre-mixed laminar flame several zones can be identified (Figure 4.4a). The gases exiting the expansion chamber and entering the burner form a *pre-heating zone* where the heat of the exiting flame causes subsequent ignition. Surrounding this zone is the luminous *primary reaction zone* whose outer surface forms the flame front. This is the hottest part of the flame. The primary reaction zone is extremely thin (0.01–0.1 mm) and thus there is not enough time available for the flame gases to complete reaction and reach equilibrium. There is a great deal of light emission from this region, which leads to an intense background signal and high noise levels, therefore negating its use for AAS. This situation changes as the partly reacted gases and free radicals pass into the *inter-conal zone*. In this region high temperatures are produced by exothermic radical combination reactions. In a typical flame, the gases will be made up of molecules such as CO, H_2, CO_2, H_2O, O_2, and N_2, and radicals such as, H, OH, C_2, CH, and CN. This is the most analytically useful part of the flame for AAS. Air entrainment into the flame leads to the formation of the *secondary reaction zone*.

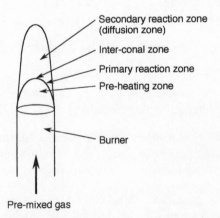

Figure 4.4a   Schematic diagram of a pre-mixed laminar flame

This zone involves reactions of the hot, partly combusted gases of the inter-conal zone and the surrounding air. It is these reactions that give rise to the characteristic blue colour of the flame.

Two flames are normally used in AAS, namely the air–acetylene flame and the nitrous oxide–acetylene flame. Both are located in a slot burner which is positioned in the lightpath of the HCL (Figure 4.4b). The choice of flame is straightforward; the air–acetylene flame (slot length 100 mm) is the most commonly used, whereas the nitrous oxide–acetylene flame (slot length 50 mm) is reserved for the more

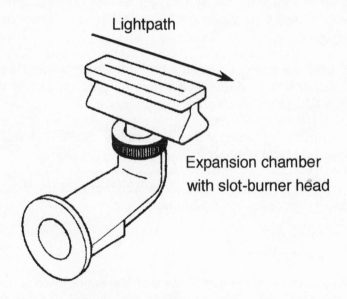

Figure 4.4b Flame-burner head

refractory elements, e.g. Al. The difference in the slot length is due to the higher burning velocity of the nitrous oxide–acetylene flame, which thus necessitates a shorter slot length to prevent flash-back into the expansion chamber. The choice of the nitrous oxide–acetylene flame for the more refractory elements may well indicate the main characteristic difference of the flames, i.e. temperature. The nitrous oxide–acetylene flame is the hotter flame (3150 K), in comparison to the cooler air–acetylene flame (2500 K).

∏ For a particular flame oxidant–fuel mixture, is it possible to vary the flame temperature?

While their general temperatures provide a general difference between the flames, it is possible to vary the temperatures slightly by altering the flame composition between a fuel-rich, a stoichiometric or a fuel-lean flame. The stoichiometric flame usually provides the highest temperature in the inter-conal zone.

The major limitations of flame AAS are as follows:

● the sample introduction system, e.g. nebuliser/expansion chamber, which is used is inefficient and requires large volumes of aqueous sample;

● the residence time, i.e. the length of time that the atom is present in the flame, is limited due to the high burning velocity of the gases, thus leading to rather high detection limits;

● an inability to analyse solid samples directly (solids require dissolution prior to analysis).

This led to a search for alternative methods of both sample introduction and sites for atomisation.

4.4.2   Graphite Furnace

One way to increase the sensitivity is to introduce discrete amounts of sample. This can be achieved by using the conventional nebuliser/expansion chamber arrangement, modified with the inclusion of a flow injection/chromatography arrangement, or as was originally carried out, by using a graphite furnace atomiser. The graphite atomiser replaces the flame-burner arrangement in the atomic absorption spectrometer.

The design of the graphite furnace or electrothermal atomiser has undergone various changes since its inception in 1961 by L'Vov. The principal of operation is that a small discrete sample (5–100 µl) is introduced on to the inner surface of a graphite tube through a small opening (Figure 4.4c). The graphite tube is arranged so that light from

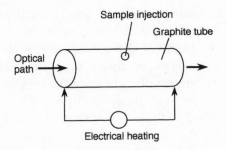

Figure 4.4c Graphite tube

the HCL can pass directly through the unit. The tube is 3–5 cm long, with a diameter of 3–8 mm. Various forms of graphite are used, including pyrolytic graphite (formed by heating the graphite tube in a methane atmosphere). The advantage of pyrolytic graphite is that it has a low gas permeability and is reasonably resistant to chemical attack. This is important as the pyrolytic graphite surface offers a more dense, impervious surface than an uncoated tube; this can reduce the tendency for carbide formation, e.g. tungsten. Heating of the graphite tube is achieved by the passage of an electric current through the unit via water-cooled contacts at each end of the tube. Careful control of the heating allows various steps to be incorporated into the programmable heating cycle. Various stages of heating (Figure 4.4d) are required to dry the sample, remove the sample matrix and finally to atomise the analyte. An additional heating cycle may be introduced for cleaning, i.e. removal of any residual material. It is the manner of these heating cycles that is the key to the success of this technique.

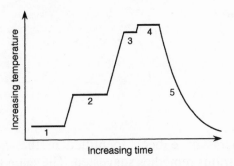

Figure 4.4d Temperature–time profile for a graphite furnace: (1) drying; (2) thermal pre-treatment (ashing); (3) atomisation; (4) cleaning; (5) cooling

The drying stage is necessary to remove any residual solvent from the sample. This can be achieved by maintaining the heat of the graphite tube above the boiling point of the solvent, e.g. water at 110°C for 30 s. The second step concerns the destruction of the sample matrix in a process called ashing; this involves heating the tube between 350 and 1200°C for ca. 45 s. In an ideal situation, the organic matrix components are removed without any loss of the analyte of interest. As can be appreciated, this is not always possible, with this being one of the major disadvantages of the technique. Various remedies for matrix modification have been evaluated and will be discussed later (see Section 4.8.2). Finally, the temperature of the graphite tube is raised to between 2000 and 3000°C for 2–3 s, allowing atomisation of the analyte of interest. It is only during this final atomisation step that the absorption of the radiation source by the atomic vapour is measured. It is common practice to have an internal flow of an inert gas (N_2 or Ar) during the drying and ashing stages in order to remove any extraneous material.

SAQ 4.4a   Would you think it necessary to operate with a gas flow during the atomisation step?

Various problems have been identified in the use of the graphite furnace and various remedies suggested. The main problem is that the analyte atoms which are volatilised from the tube wall come into contact with the cooler gas. This is because an electric current, i.e. the method of heating, is applied directly to the tube. Therefore, it takes longer for the gas contained within the tube to be heated and so there is a temperature lag. One way to overcome this problem would be to delay the heating of the sample until a temperature equilibrium has been established. A solution was proposed by L'Vov, who incorporated a pyrolytic graphite platform within the graphite tube (Figure 4.4e). In this situation, the sample is deposited on to the L'Vov platform where heating occurs primarily by radiation from the tube walls, thereby inducing a temperature lag. The platform then reaches the atomisation temperature when the tube wall and the gas have reached equilibrium (Figure 4.4f). In addition, the atoms are volatilised into the gaseous state at a higher temperature, thereby reducing potential chemical interferences.

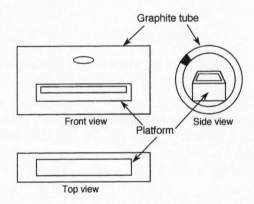

Figure 4.4e Schematic diagram of the L'Vov platform

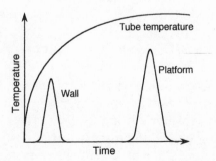

Figure 4.4f Comparison of wall and platform atomisation

A different approach to the problem has been proposed more recently. In this case, the sample is placed directly on to a probe which is inserted into the graphite furnace in order to facilitate drying and ashing. However, prior to atomisation the analyte-containing probe is removed from the graphite furnace. Once the atomiser has reached the pre-set atomisation temperature, the probe is reintroduced into the tube and atomisation occurs under isothermal conditions and the absorption measurement made. Probe atomisation has several disadvantages. The continuous introduction and withdrawal of the probe can lead to mechanical and thermal stress, with a consequent reduction in performance and lifetime of the probe. In addition, the reintroduction of the probe into the hot environment of the graphite tube can entrain air which could also adversely effect the tube lifetime.

4.4.3   Hydride Generation and Cold Vapour Techniques

Hydride generation is a special form of sample introduction which is exclusively reserved for a limited number of elements that are capable of forming volatile hydrides (e.g. As, Bi, Sb, Se, and Sn). In this situation, an acidified sample solution is reacted with a sodium tetraborohydride solution, and after a short time, the gaseous hydride is liberated. An equation to describe the generation of the arsine hydride (AsH_3) can be written as follows:

$$3BH_4^- + 3H^+ + 4H_3AsO_3 \longrightarrow 3H_3BO_3 + 4AsH_3 + 3H_2O \quad (4.4)$$

However, when a basic borohydride is added to an acidic solution, excess hydrogen is liberated:

$$BH_4^- + 3H_2O + H^+ \longrightarrow H_3BO_3 + 4H_2 \quad (4.5)$$

Various gas–liquid separation devices have been used for either batch or on-line separation. However, whichever separation device is used the improvement in sensitivity is always high. The generated hydride is transported to an atom cell by using a carrier gas. The cell used for atomisation consists of either an electrically heated or flame heated quartz tube.

SAQ 4.4b	Why does the atom cell require heating in hydride generation?

Cold vapour generation is exclusively reserved for the element mercury. In this situation, the mercury present in the sample is reduced, usually by using tin(II) chloride, to elemental mercury, as follows:

$$Sn^{2+} + Hg^{2+} \longrightarrow Sn^{4+} + Hg^O \tag{4.6}$$

The mercury vapour thus generated is then transported by a carrier gas to the atom cell; the latter takes the form of a long-path glass absorption cell which is located in the path of the HCL. Mercury is monitored at 253.7 nm.

∏ Why is it sensible to use a long-path absorption cell?

The longer the pathlength, then the greater will be the sensitivity (Beer–Lambert law; see Section 4.2).

4.5 SAMPLE INTRODUCTION

The introduction of an aqueous sample into the flames used in atomic absorption spectroscopy is almost exclusive to the use of a nebuliser/expansion chamber arrangement.

4.5.1 Nebuliser/Expansion Chamber

The pneumatic nebuliser consists of a concentric stainless steel tube into which a Pt/Ir capillary tube is fitted. The sample is drawn up through the capillary by the action of the oxidant gas (air) escaping through the exit orifice that is located between the outside of the capillary tube and the inside of the stainless steel tube. The action of the escaping air and liquid sample is sufficient to break up the latter into a coarse aerosol; this is known as the Venturi effect. The typical uptake rate of the nebuliser is between 3 and 6 ml min^{-1} (see also Section 5.3.1 which describes the pneumatic nebuliser used in atomic emission spectroscopy (AES)).

SAQ 4.5   Compare the pneumatic nebulisers used for flame atomic absorption and inductively coupled plasma atomic emission spectroscopy.

The expansion chamber has a dual function. The first is to convert the aqueous sample solution into a coarse aerosol using the oxidant gas, which then allows this aerosol to be dispersed into a finer form for transport to the burner for atomisation, or to allow residual aerosol particles to condense and go to waste. Secondly, the arrangement also facilitates safe pre-mixing of the oxidant and fuel gases in the expansion chamber prior to their introduction into the laminar flow burner. Figure 4.5 shows a schematic diagram of a typical arrangement.

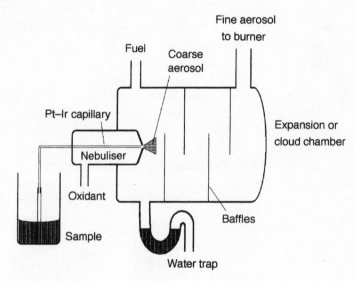

Figure 4.5 Nebuliser/expansion chamber arrangement

4.6 WAVELENGTH SELECTION AND DETECTION

As the selectivity of AAS is achieved by the 'lock and key' effect (described earlier) the role of the spectrometer is not as critical as in AES (see Section 5.4). Nevertheless, the spectrometer still has an important focus for AAS. The optical arrangement used in the spectrometer is the Czerny–Turner configuration, and a full description of this is given in Section 5.4.1. It should be noted, however, that the Czerny–Turner monochromator used in AAS has a focal length of only 0.25–0.5 m, with a grating containing only 600 lines per mm and a relatively poor resolution of 0.2–0.02 nm. Of greater concern in AAS is the optical arrangement prior to the spectrometer. Two configurations are possible, namely single beam and double beam. In a single-beam instrument (Figure 4.6a), light from the hollow-cathode lamp is focused on to the atomic vapour generated in the atomiser, e.g. the flame, and then passes to the monochromator. In the double-beam arrangement (Figure 4.6b), light from the lamp is split by a mirrored chopper. This allows half of the light to pass through the atomiser while the remaining half is diverted. The two beams are then recombined by a half-silvered mirror and then pass to the monochromator. The double-beam instrument does offer some advantages as it allows the correction of any HCL fluctuations caused by warm-up, drift and source noise, thus leading to improved precision in the absorbance measurements.

The attenuation of the hollow-cathode lamp radiation by the atomic vapour is detected by a photomultiplier tube (PMT), and a full description of the operation of this can be found in Section 5.4.3. For the sake of completeness, however, a brief outline is presented here. The PMT is a device for proportionally converting photons to current. The incident light strikes a photosensitive material which converts the light into electrons (the photoelectric effect). The generated electrons

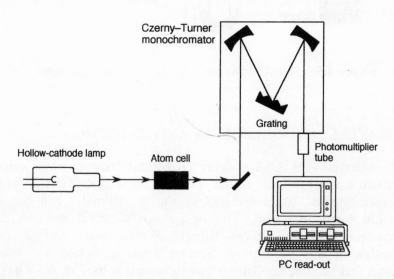

Figure 4.6a Atomic absorption spectrometer: arrangement for a single-beam instrument

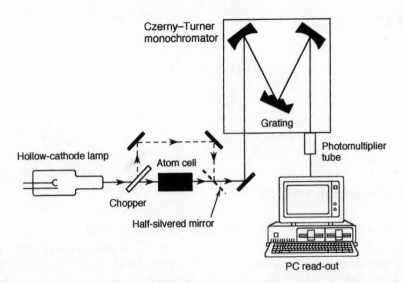

Figure 4.6b Atomic absorption spectrometer: arrangement for a double-beam instrument

are then focused and multiplied by a series of dynodes prior to collection at the anode. These multiplied electrons (i.e. electrical current) are then converted into a voltage signal which is then transferred via an analog-to-digital (A/D) converter to a suitable computer for processing purposes.

4.7 BACKGROUND CORRECTION METHODS

The occurrence of absorbance and scatter from molecular species can be overcome by the use of various background correction methods. The most common of these are described below.

| SAQ 4.7 | Why should the presence of molecular species be a problem in AAS? |

SAQ 4.7

4.7.1 Continuum Source

As the name suggests, this method of background correction is facilitated by the use of a continuum source, e.g. a deuterium lamp. In the atomisation cell, absorption is possible from both atomic and molecular species. By measuring the absorption that occurs from the line source (HCL) and comparing it with the absorbance that results from the continuum source, a corrected absorption signal can be obtained. This is because the atomic species absorb the specific radiation associated with the HCL source, whereas the absorption of radiation by the continuum source for the same atomic species will be negligible. Instrumentally, this involves the alignment of the light from both the hollow cathode and the deuterium source through the flame to the spectrometer/detector. This is achieved in practice by a beam splitter in the form of a rotating mechanical chopper. However, the use of a beam splitter can lead to a loss in intensity of the light which reaches the detector. In addition, great care is needed to align both the hollow cathode and deuterium sources along the same path in the atomisation cell. Such disadvantages of this correction process have therefore led to the search for alternative methods.

4.7.2 Smith–Hieftje

The disadvantages of the continuum source correction technique can be alleviated by the Smith–Hieftje method of background correction. In this approach, a single HCL is used, which is operated under both high and low (or normal) current conditions. Operating the HCL at a high current induces self-reversal in the emission profile of the lamp (Figure 4.7a), and it is this feature which is exploited in the Smith–Hieftje

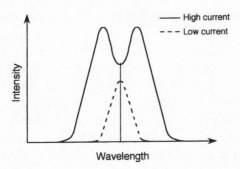

Figure 4.7a Emission profile of a hollow-cathode lamp for Smith–Hieftje background correction (note the line self-reversal at high operating current)

method. The principle is as follows. The HCL operating at a low current is able to absorb radiation from both the atomic (desired) and molecular (unwanted) species, as would normally occur. If the HCL is then subjected to a pulse of high current, self-reversal occurs, which effectively splits the profile into two components (Figure 4.7b). Incomplete self-reversal in the HCL will lead to a corresponding

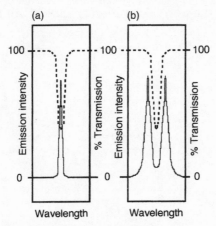

Figure 4.7b Schematic representation of the Smith–Hieftje background correction technique. (a) Low current: hollow-cathode lamp emits a sharp line. Both atomic and molecular absorptions are measured. (b) High current: hollow-cathode lamp emission profile is effectively split into two on either side of the atomic-absorption-line profile. Only broad-band molecular absorption is measured

incomplete resolving of the atomic line profile from the molecular band interferents. As the atomic species no longer coincide at exactly the same wavelength (as is the case with the HCL operating at low current) they are not seen, whereas the broad-band absorption of the molecular species are still observed. A subtraction process thus allows a corrected absorption signal to be monitored. The major advantage of this type of correction method over the continuum-based approach is that only a single lamp is required so no loss in sensitivity can occur due to the use of a beam splitter or to misalignment in the atomisation cell.

4.7.3 Zeeman Effect

The Zeeman-effect background correction method also utilises line splitting in its correction process. In this case, splitting of the atomic spectral line is achieved by the application of a strong magnetic field. The magnetic field is normally applied to the atomisation cell (graphite furnace), although it is possible to apply it to the radiation source or the HCL (for flame AAS). Thus, when an atom is placed in a magnetic field and the absorption observed with polarised light, the absorption line profile is split into two components, which are symmetrically displaced around the normal position (Figure 4.7c). For most elements,

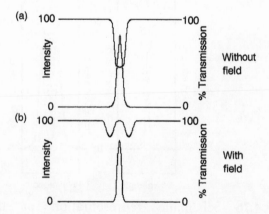

Figure 4.7c Schematic representation of the Zeeman background correction technique. (a) No magnetic field applied: normal situation in atomic absorption spectroscopy. (b) Magnetic field applied: atomic-absorption-line profile is split into its σ-components; no significant overlap with hollow-cathode lamp emission profile

therefore, the central σ-component occurs at the absorption wavelength of the element of interest, with no atomic absorption occurring at the two π-components. However, any molecular broad-band absorption that is present will be absorbed by the two π-components. By alternating between a situation in which the magnetic field is 'off', i.e. both atomic and molecular absorptions are observed, to one in which the magnetic field is 'on', i.e. only the background is observed, it is possible to obtain a corrected atomic absorption signal (Figure 4.7c). In the magnetic 'on' situation a polarising filter allows the σ-components, used to measure the background absorption, to be recorded by the detector. Various forms of the Zeeman-effect correction method have been developed commercially; these include the application of either an AC magnetic field or a constant magnetic field, with line splitting occurring by the use of a rotating polariser. This form of correction method is almost exclusively reserved for graphite-furnace work owing to the high costs involved.

4.8 INTERFERENCES

The generation of good, reliable and reproducible data is dependent upon the skill of the analyst and the utilisation of a suitable analytical technique.

∏ If the analyst is capable of utilising an instrument to its full potential, why is it that inaccurate data can still be generated?

The reason may not, at first, be obvious, but it is likely to be due to some form of interference. These may be classified in various ways, e.g. in flame AAS, physical, spectral and chemical interferences may occur, plus in addition, those resulting from various ionisation processes. Therefore, in order to obtain reliable and accurate data we require an understanding of the processes that can lead to such interferences and the remedies that can be used to alleviate them.

4.8.1 Flame Atomisation Interferences

The dominant interferences that occur in flame AAS are chemical, ionisation, physical and spectral. Their mode of occurrence and appropriate remedies are described below.

Chemical

This is the dominant type of interference in flame atomic absorption spectroscopy. Chemical interferences arise when the element to be determined forms a thermally stable compound with certain molecular or ionic species that are present in the sample solution. The best examples of this are when the presence of phosphate, silicate or aluminate in the sample solution cause a suppression of the alkaline earth metal absorption signal in the air–acetylene flame. Figure 4.8a shows the effect of increasing amounts of phosphate on the Ca absorption signal at 422.7 nm. This signal depression is due to the formation in the flame of the thermally stable compound, calcium pyrophosphate.

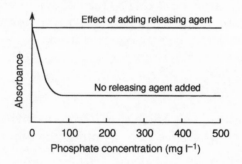

Figure 4.8a Chemical interference in flame atomic absorption spectroscopy, showing the effect of phosphate concentration on calcium absorbance

SAQ 4.8a   Can you think of any remedies to alleviate chemical interferences?

SAQ 4.8a

Ionisation

These are vapour-phase interferences and hence occur predominantly in the flame. The problem is most severe for alkali and alkaline earth metals, where their low ionisation potentials can lead to ionisation in the relatively hot environment of the flame.

∏ Why should ionisation be a problem in atomic absorption spectroscopy?

Obviously, if ionisation occurs then no atoms will be present, and therefore no absorption signal will be detected:

$$Na \longrightarrow Na^+ + e^- \qquad (4.7)$$

This non-desirable situation can be prevented by the addition of an ionisation suppressor or buffer. A typical ionisation buffer would be another alkali metal, e.g. caesium. Addition of an excess of the latter leads to its ionisation in the flame in preference to the analyte, e.g. sodium. Therefore, the addition of excess Cs suppresses the unwanted ionisation of Na. This effect is called the 'mass action' effect (Figure 4.8b).

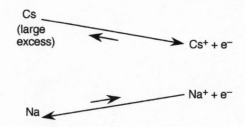

Figure 4.8b Ionisation interference in flame atomic absorption spectroscopy, showing the effect of excess caesium on the determination of sodium (mass-action effect)

Physical

Physical interferences in flame atomic absorption spectroscopy are related to the transport of the sample solution to the flame. Hence this type of interference is related to the nature of the sample solution, e.g. its viscosity and its conversion into an aerosol by the spray chamber. The aerosol which is formed is dependent upon the surface tension, density and viscosity of the sample solution. Interferences of this type can be controlled by careful matrix matching of sample and standard solutions. If this is not possible, the analyst must then use the method of standard additions (see Section 1.3) to alleviate any potential interferences. To successfully utilise the standard additions method it is essential that the calibration is linear over the absorbance range to be covered and that a reasonable number of solutions (five or six) are prepared. By plotting an aqueous calibration plot alongside the standard additions plot it is possible to visually observe the physical matrix effect. If the two plots are not parallel, then some physical matrix interference is occurring with either an enhanced or suppressive effect.

SAQ 4.8b	Explain how you would prepare a standard additions plot for the analysis of calcium in sea water.

SAQ 4.8b

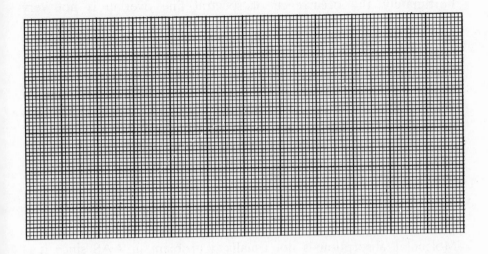

Spectral

Fortunately, the occurrence of spectral line overlap is not very common due to the high selectivity of AAS offered by the 'lock and key' effect. However, some well known examples do occur, e.g. the resonance line of copper (324.754 nm) has a line coincidence with europium (324.753 nm). However, unless the Eu is in excess compared to the Cu (ca. 1000 times) no significant interference with the Cu determination will occur. In addition to atomic spectral overlap, molecular band absorptions can coincide. Examples of these are calcium hydroxide which has an absorption band on the barium wavelength of 553.55 nm, and lead at 217.0 nm which has a molecular absorption from sodium chloride.

∏ How can you correct for molecular absorption in AAS?

Molecular absorption is not usually a problem in AAS since it is possible to correct by using various background correction techniques (See Section 4.7).

4.8.2 Graphite Furnace Interferences

The main types of interference problems in graphite furnace atomic absorption spectroscopy are those due to background absorption and scattering effects. Light scattering in the graphite tube can result from mist and smoke from particles that form at the cooler ends of the tube. This can lead to the detrimental occurrence of molecular absorption. Background absorption is particularly troublesome from alkaline or alkaline earth halides, e.g. sodium chloride. It is possible to correct for molecular absorption effects by using background correction methods (see Section 4.7).

An alternative source of interference in graphite furnace AAS is that due to memory effects.

∏ Why do memory effects occur in a graphite furnace?

In this situation, residual analyte from incomplete atomisation or incomplete tube cleaning from the previous analysis can lead to an enhancement of the signal response in later analyses. This is particularly troublesome for those elements which form refractory oxides or carbides. The remedy is to increase the atomisation temperatures (and time) and/or cleaning temperatures (and time).

Physical

Physical interferences are not so important in graphite furnace atomisation, although the viscosity and surface tension of the sample solution can lead to problems in the pipetting of microlitre samples. The problem is made worse if the sample spreads after application to the graphite furnace, whereby pre-atomisation of the analyte may occur. This problem is partly remedied by using peak-area measurements.

∏ How might you prevent sample spreading in the graphite furnace?

This may be achieved by placing physical barriers in the graphite tube to prevent the sample from spreading. Typically, this is applicable to the L'Vov platform which has a limited capacity in which to place the aqueous sample.

Chemical

Chemical interferences in graphite furnace AAS can be classified into two types: (i) the formation of volatile compounds which are lost prior to atomisation; (ii) the formation of a more stable compound which is not totally atomised. In the first situation, it may be that a volatile compound is formed *in situ* by the effect of heat on the matrix and/or solvent, or that the analyte may be present in the sample matrix in a volatile form. Thermal pre-treatment/atomisation curves are useful in these circumstances as they allow the best thermal treatment conditions to be identified (Figure 4.8c). Volatile-compound formation can be prevented by the use of matrix modification. In this method, the addition of a suitable compound(s) allows volatile compounds to be retained until the atomisation step has been reached.

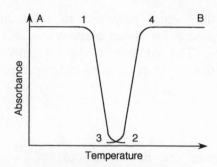

Figure 4.8c Thermal pre-treatment/atomization curves, where curve A represents the absorbance measured at the optimum atomisation temperature as a function of the ashing temperature, and B is the absorbance versus the atomisation temperature: (1) maximum thermal pre-treatment temperature; (2) lowest temperature at which the analyte is quantitatively volatilised; (3) appearance temperature of the analyte; (4) optimum atomisation temperature

The occurrence of a more stable compound can also have a detrimental effect on the signal response. In this situation, however, it is not that the analyte has been pre-atomised, but that atomisation is only partly successful. The situation is best illustrated by elements that form carbides.

SAQ 4.8c   Can you suggest any elements that form carbides?

A reduction in carbide formation can be achieved if the pyrolytic graphite surface is maintained in good condition (and replaced if deterioration is observed), and adequate cleaning procedures are adopted. In addition, carbide formation can be eliminated if the graphite surface is replaced with a metal atomiser, e.g. tungsten or tantalum, or by coating the graphite tube with tantalum.

Matrix Modification

Matrix modification is the term used to describe the process of stabilising the analyte until the atomisation stage is reached (see Figure 4.4d), or to enable the matrix components to be volatilised during the ashing stage (Figure 4.4d) in graphite furnace AAS. The aim of matrix modification is to establish a different thermal separation between the matrix and the analyte components. In practice, this involves the addition of a reagent to the sample or standard solution in a large excess. For example, the addition of ammonium nitrate eliminates interferences due to sodium chloride (a potential interference in lead determinations; see Section 4.8.1) according to the following equation:

$$NH_4NO_3 \quad + \quad NaCl \quad = \quad NaNO_3 \quad + \quad NH_4Cl \qquad (4.8)$$

decomposes m.p. 1079 K; decomposes sublimes at 618 K;
  at 483 K   b.p. 1691 K   at 653 K   b.p. 798 K

In this situation, the products formed (sodium nitrate and ammonium chloride) are removed by decomposition or sublimation below 700 K. In addition, any excess ammonium nitrate is easily removed during the pre-treatment (ashing) stage.

The loss of arsenic, selenium and tellurium can be prevented by the addition of excess nickel (Figure 4.8d). This is presumably due to the formation of thermally stable compounds, such as nickel arsenide, etc. In this situation, arsenic can be pre-treated up to temperatures of 1700 K. Other common matrix modifiers include magnesium nitrate and salts of palladium and platinum.

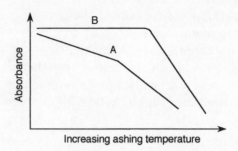

Figure 4.8d Matrix modification of selenium using nickel: (A) 2 ng Se; (B) A + 20 μg Ni

Summary

The basis of atomic absorption as a quantitative technique for obtaining analytical data is discussed. Details of instrumentation are highlighted and discussed in the context of sample introduction, measurement and signal read-out. The types of interferences that can lead to erroneous analytical measurements are described and appropriate remedies suggested.

Objectives

On completion of this chapter you should be able to:

- describe the operation of a hollow-cathode lamp;

- outline why a hollow-cathode lamp is required for AAS;

- suggest, with reasons, why other radiation sources may be used in AAS;

- describe the common flames used for AAS;

- describe the chemical processes that occur in a flame;

- describe the flame structure;

- suggest alternatives to the flame atomisation cell;

- describe the mode of operation of a graphite furnace;

- write chemical equations for the formation of hydrides and elemental mercury;

- describe the sample introduction strategy for flame AAS;

- describe why a double-beam arrangement is better than a single-beam arrangement in AAS;

- outline the different background correction techniques used in AAS;

- appreciate the limitations of each of the background correction techniques;

- appreciate the different interferences that can occur in flame and graphite furnace AAS;

- outline the use of matrix modification in graphite furnace AAS.

5. Atomic Emission Spectroscopy

The instrumentation for atomic emission spectroscopy (AES) consists of an atom cell, spectrometer/detector and read-out device. An ideal excitation source for AES would have the following features:

(i)   high selectivity;

(ii) high sensitivity;

(iii) high accuracy;

(iv) high precision;

(v)   capacity for multi-element determinations;

(vi) freedom from matrix interferences.

∏ How would you prioritise these ideal features?

All of these are important and are necessary in any multi-element technique. We will see in this chapter how close plasma sources come to possessing all of these features.

The classical AES sources, such as the arc and the spark, all suffer from poor stability, low reproducibility, and substantial matrix interferences. However, the plasma-based sources can offer most of the benefits of an ideal AES source.

5.1 FLAME EMISSION SPECTROSCOPY

Flame emission spectroscopy (FES) is probably the oldest of the spectroscopic techniques. Early work involved the introduction of metal salts into flames and visual observation of the colours that were produced.

SAQ 5.1   The introduction of a sodium salt into a Bunsen burner is characterised by the colour of the flame. What colour would you expect the flame to be?

Since this early work, more modern instruments have replaced the mode of operation so that quantitative analysis is possible. A typical FES instrument has the same design as that used for conventional flame AAS, which was previously described in Section 4.4.1. In this situation, no instrument modification is necessary and no hollow-cathode lamp is required. In contrast to flame AAS, where the specificity of the technique is derived from the use of the hollow-cathode lamp, no such advantage is gained in FES. This means that spectral interferences cannot be resolved by the relatively poor resolution of the monochromator. Therefore, all of the sample that enters the flame is potentially available for detection. The relatively low temperature of the flame, even when using a nitrous oxide–acetylene mixture, means that potential interferences from molecular species, which are principally derived from the flame gases, can cause erroneous results to be obtained.

An analogous technique that uses a much simpler arrangement is flame photometry (FP). This technique is almost exclusively reserved for alkali metals because of their low excitation potentials (e.g. sodium, 5.14 and potassium, 4.34 eV) and hence their ability to be easily ionised. This allows a cooler flame to be used in conjunction with a simpler spectrometer. Typically, the flame may consist of air–propane, air–butane or air–natural gas, while the spectrometer is basically an interference filter. The cool nature of the flame prevents other metals from being excited, and so the method is relatively free of interferences. As a monochromator is not necessary, the use of an interference filter allows a large excess of light emission to be viewed by the detector. The expensive photomultiplier tube can therefore be replaced by a significantly cheaper photodiode or photoemissive detector. Thus, a simple, robust and inexpensive instrument is available for the determination of, for example, potassium (766.5 nm) or sodium (589.0 nm) in clinical or environmental samples.

5.1.1 Limitations

There are four types of interference that can occur in flame atomic absorption spectroscopy (see Section 4.8); all of these can also occur in FES. They are as follows:

chemical interferences;

ionisation interferences;

physical interferences;

spectral interferences.

5.2 PLASMA SOURCES

A plasma can be thought of as the co-existence in a confined space of the positive ions, electrons and neutral species of an inert gas, typically argon or helium. Classification of the common plasma sources is made according to the method of inputting power to the gas. For example, the inductively coupled plasma (ICP) is sometimes referred to as the radiofrequency ICP; other types of plasma are the direct current plasma (DCP) and microwave-induced plasma (MIP). The subject of plasma temperature still remains an area for discussion.

∏ How do you measure temperature?

This is probably more difficult to answer than you might initially think, and particularly so in the context of plasmas. In the laboratory, temperature is measured by using a thermometer. However, in a plasma source this is not possible.

Even though the plasma is electrically neutral, it is not in thermodynamic equilibrium. Hence it is not possible to characterise a single temperature. Four temperatures can be used to characterise the plasma, namely excitation, ionisation, electron and gas temperatures. The excitation temperature is a measure of the population density of the energy levels (see Section 3.4), the ionisation temperature is a measure of the population density of the different ionisation states, the electron temperature is a measure of the kinetic energy of the electrons, and the gas temperature is a measure of the kinetic energy of the atoms. In each case, the typically quoted plasma temperatures range from 7000–10000 K (using the ICP as an example here). Unfortunately, this is well above any direct form of measurement of which you will probably be aware, i.e. a thermometer, thermocouple, etc. Therefore, in order to measure the temperature of a plasma source the scientist must use alternative approaches, such as spectroscopic methods. For example, the excitation temperature can be measured by means of the Boltzmann equation (see Section 3.5). The plasma temperature is also inhomogenous, i.e. there is a variation in temperature, which occurs both radially and axially. For further details on plasma temperature measurement the interested reader should refer to the section for further reading given later in this book (Section 8.2).

5.2.1 Inductively Coupled Plasma

An ICP is formed within the confines of three concentric glass tubes or plasma torch (Figure 5.2a). Each tube has an entry point, with those of the intermediate (plasma) and external (coolant) tubes being arranged tangentially to that of the inner tube, where the latter consists of a capillary tube through which the aerosol is introduced from the nebulisation/spray chamber. Located around the outer glass tube is a coil of copper tubing through which water is recirculated. The power input to the ICP is achieved through this load or induction coil, and is typically in the range 0.5–1.5 kW at a frequency of 27 or 40 MHz. The inputted power induces an oscillating magnetic field, whose lines of force are axially orientated inside the plasma torch and follow elliptical

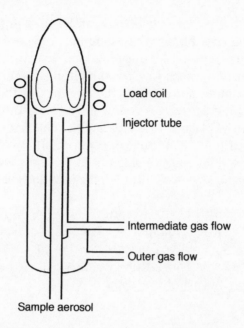

Figure 5.2a Schematic diagram of an inductively coupled plasma torch

paths outside the induction coil (Figure 5.2b). At this point in time, no plasma exists. In order to initiate the plasma, the carrier gas flow is first switched off and a spark is then provided momentarily from a Tesla coil, which is attached to the outside of the plasma torch by means of a piece of copper wire. Instantaneously, the spark, which is a source of 'seed' electrons, causes ionisation of the argon carrier gas. This process is self-sustaining, so that argon, argon ions and electrons now co-exist within

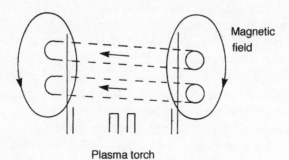

Figure 5.2b Formation of an inductively coupled plasma

the confines of the plasma torch, and can be seen protruding from the top in the shape of a bright white luminous bullet. This characteristic bullet shape is formed by the escaping high-velocity argon gas causing the entrainment of air back towards the plasma torch itself. In order to introduce the sample aerosol into the confines of the hot plasma gas (7000–10 000 K) the carrier gas is now introduced; this 'punches' a hole in the centre of the plasma, thus creating the characteristic doughnut or toroidal shape of the ICP. In the conventional ICP system, the emitted radiation is viewed laterally, or side-on. In this way, the radiation of the element of interest is 'viewed' through the luminous plasma.

SAQ 5.2a

> Can you identify any potential difficulty in viewing the luminous plasma side-on? Can you suggest an alternative viewing position?

Figure 5.2c compared the typical background emission characteristics observed for a conventional, side-on viewed plasma with that of an axially viewed plasma. The two major features of the spectra are the presence of a large number of emission lines and a background continuum. The emission lines are mainly due to the source gas (argon), but also result from the presence of atmospheric gases, e.g. N_2, and the breakdown components of water, e.g. OH. It can be seen that the side-on viewed plasma has a higher background continuum than the axially viewed plasma. In both cases, the background continuum is due to radiative recombination of electrons and ions $(M^+ + e^- \longrightarrow M + h\nu)$ and the radiation loss of energy by accelerated electrons (bremsstrahlung radiation). Commercial instruments are now available that allow axial viewing of the plasma.

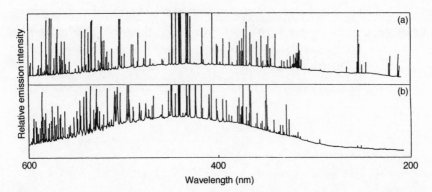

Figure 5.2c Comparison of spectral features and background between (a) an axially viewed ICP, and (b) a conventional side-on viewed ICP

5.2.2 Direct Current Plasma

The DCP is essentially an electrical discharge which is struck between two anodes and a cathode. The arrangement of the electrodes (Figure 5.2d) is such that tangentially flowing argon gas from the two anodes forms an inverted V-shaped plasma. The cathode is located directly above the central apex.

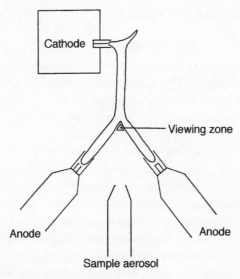

Figure 5.2d Direct current plasma

SAQ 5.2b	Do you think that the position of the cathode directly above the two anodes offers any advantage for the DCP?

The sample aerosol is directed into the apex of the V, but it does not penetrate the plasma. This can lead to spectrochemical interferences, as the sample aerosol does not experience the same temperature environment as in the ICP.

5.2.3 Microwave-induced Plasma

This is probably the least common type of plasma source used for atomic emission spectroscopy. While some work has focused on its use for solution introduction by using a high-power source (1 kW), it is as a detector for gas chromatography that the microwave-induced plasma (MIP) has achieved a niche market.

This plasma is usually generated in a disc-shaped cavity resonator, e.g. a Beenakker-type resonator. Located through the cavity is the capillary tube, ca. 1–2 mm internal diameter, in which the plasma will be formed. Carrier gas is introduced through the capillary tube prior to initiation of the plasma. Typical carrier gases for a MIP include argon and nitrogen, but more commonly helium. Power to the cavity is supplied by a microwave generator, which provides 50–200 W at 2.45 GHz. Initiation of the plasma, as for the ICP, requires the addition of a spark from a Tesla coil.

The MIP is characterised by a high excitation temperature (7000–9000 K), but a low gas temperature (1000 K). This situation favours the atomic emission of non-metals.

The analysis of non-metals by using the MIP has been extensively developed, such that a coupled gas chromatography–MIP–AES system is now commercially available (Figure 5.2e).

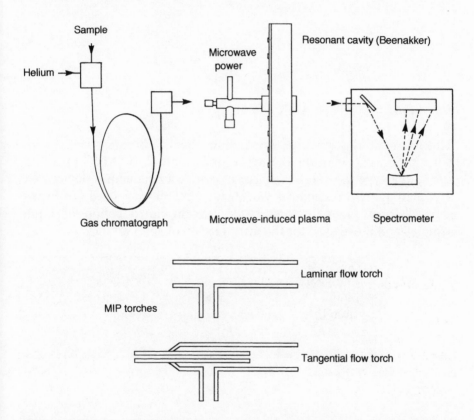

Figure 5.2e Coupled gas chromatography–microwave-induced plasma–atomic emission spectroscopy system (plus examples of typical MIP torches)

SAQ 5.2c   Would the ICP be a good source for the analysis of non-metals?

SAQ 5.2c

5.3 SAMPLE INTRODUCTION

This has been described as the Achilles' heel of atomic spectroscopy. Since the inception of the plasma source as a tool for AES in the mid-1960s, the type and performance of sample introduction devices has occupied analytical chemists worldwide. The devices used are of two general forms, namely those used for the introduction of liquid samples and those used for the introduction of solid samples.

SAQ 5.3a

Sample introduction in AES requires the conversion of the sample into a (dry or wet) aerosol. Why should the use of an aerosol be an effective means of introducing the sample?

5.3.1 Nebulisers

The most common method of liquid sample introduction in AES involves the use of a nebuliser. The type of nebuliser used in modern instruments has not altered significantly since it was first introduced, despite of its inefficiency. Most of the nebulisers available, with the exception of the ultrasonic nebuliser, have transport efficiencies of between 1 and 2%. Transport efficiency is defined as the amount of the original sample solution that is converted to an aerosol and then into the plasma source. The basic function of the nebuliser is to convert an aqueous sample into an aerosol by the action of a carrier gas. In order to produce an aerosol of sufficient particle size (ideally < 10 μm), thus avoiding substantial cooling/extinguishing of the plasma, it is necessary to introduce the generated aerosol into a spray chamber. The latter will further reduce the original aerosol particle size towards the ideal size by providing a surface for collisions and/or condensation. The generation of condensation, however, contributes towards the inefficiency of the nebuliser/spray chamber sample introduction system. The nebulisation of aqueous samples is affected by their viscosity and surface tension, both of which have an influence on the carrier gas and uptake rate. It is possible, however, to overcome these problems by using a peristaltic pump to transport the aqueous sample to the nebuliser.

The various types of nebuliser that are commonly used are described below.

Pneumatic Concentric Nebuliser

This is probably the most common type of nebuliser used today. It consists of a concentric glass tube through which a capillary tube is fitted (Figure 5.3a). The sample is drawn up through the capillary by

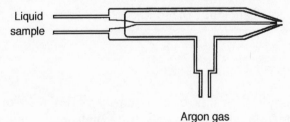

Argon gas

Figure 5.3a Schematic diagram of the pneumatic concentric nebuliser

the action of the carrier gas (argon, as a pressure of 70–350 kPa) escaping through the exit orifice that exists between the outside of the capillary tube and the inside of the glass concentric tube. The size of the toroid created by the two tubes is of the order of 10–35 μm in diameter. The action of the escaping argon gas and liquid sample is sufficient to produce a coarse aerosol (the Venturi effect). The typical uptake rate of this nebuliser is between 0.5 and 4 ml min^{-1}. As a consequence of the Venturi effect it is not necessary to use a peristaltic pump.

Cross-flow nebuliser

In this nebuliser, two capillary needles are positioned at 90° to each other, with their tips not quite touching (Figure 5.3b). The carrier gas flows through one capillary tube, while the liquid sample is pumped through the other capillary. At the exit point, the force of the escaping carrier gas is sufficient to shatter the sample into a coarse aerosol.

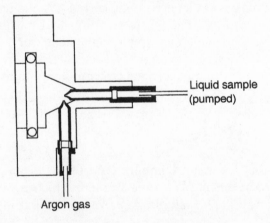

Liquid sample
(pumped)

Argon gas

Figure 5.3b   Schematic diagram of the cross-flow nebuliser

High-solids Nebuliser

As its name implies, this type of nebuliser is designed to generate coarse aerosols from aqueous samples with high-solids contents (up to 20%), the latter often being termed as slurries. The expression 'high-solids nebuliser' is often used to describe a whole range of designs that are both commercially and non-commercially available. The designs of the different nebulisers that are available can probably be best described if we consider the ubiquitously named V-groove nebuliser as a model (Figure 5.3c). In this case, the sample solution is pumped along a V-grooved channel; midway along this channel is a small orifice through which the carrier gas can escape. As the sample passes over the orifice the escaping gas causes the production of a coarse aerosol.

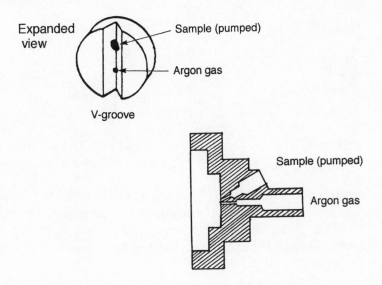

Figure 5.3c   Schematic diagram of the V-groove, high-solids nebuliser

Ultrasonic Nebuliser

In an ultrasonic nebuliser the sample is delivered on to a vibrating piezoelectric crystal (at a frequency of 200–10 MHz). The action of the vibrating crystal is sufficient to transform the liquid sample into an aerosol, which is then transported by the argon carrier gas through a heated tube and then a condenser (Figure 5.3d). This has the effect of removing the solvent. In this situation, the aerosol is desolvated and reaches the plasma source as a fine, dry aerosol. The major advantage

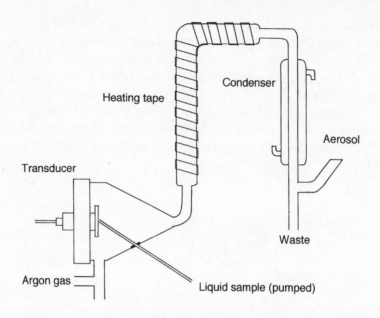

Figure 5.3d   Schematic diagram of the ultrasonic nebuliser

of the ultrasonic nebuliser is its increased transport efficiency, which is of the order of 10%, when compared to the pneumatic nebuliser. However, although many variations of the ultrasonic nebuliser have been introduced since the mid-1960s, most of these have not proved to be very successful.

5.3.2   Spray Chambers and Desolvation Systems

If introduced into the plasma source directly, the coarse aerosols generated by the nebulisers described above would either extinguish or induce cooling of the plasma. This in turn would lead to severe matrix interferences unless a spray chamber is added to the system. Several common types exist (Figure 5.3e) with all of these serving a common purpose, i.e. to further reduce the aerosol towards an ideal particle size. It has been determined that this ideal size for desolvation and ionisation/excitation in the plasma source is ca. 10 μm. The most common of these is known as the Scott double-pass spray chamber.

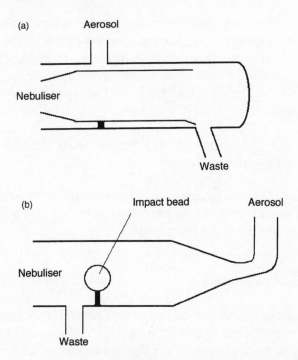

Figure 5.3e Types of spray chamber: (a) Scott double-pass type; (b) impact-bead spray chamber

∏ The spray chamber acts as an aerosol 'sorter'. What features would be ideal for a spray chamber?

An ideal spray chamber should have all of the following features:

- a large surface area to induce collisions and fragmentation of the coarse aerosol;

- minimal dead volume to prevent dilution of the sample;

- easy removal of condensed sample to waste without inducing pressure pulsing.

5.3.3 Discrete Sample Introduction

In this case, the sample is introduced into the plasma source as a discrete 'unit' of sample interspersed between either the carrier gas or a liquid carrier stream. Discrete sample introduction has the advantage of presenting the plasma source with all of the analyte in a short period of time. Therefore, while the residence time of the analyte within the plasma is short, all of the analyte is available for analysis. This has the obvious benefit that improved analyte sensitivity should be achievable.

Π What type of sample would benefit from the use of discrete sample introduction devices?

Discrete sample introduction devices are particularly beneficial when the sample is only available in limited quantities, e.g. biological tissue. Perhaps you can think of other sample types?

Electrothermal Vaporisation

The principle of electrothermal vaporisation (ETV) is that the sample should be vaporised and not atomised (in contrast to electrothermal or graphite furnace atomic absorption spectroscopy). However, this is not always achievable, particularly for the more volatile elements, e.g. cadmium, where atomisation will usually occur and thus leads to a loss in transport efficiency. Various attempts have been made to remedy this situation, including the use of matrix modification or the addition of small amounts of chlorofluorocarbons to the carrier gas.

In ETV the sample (usually liquid, but it is also possible to introduce small amounts of solids) is pipetted on to a graphite or modified graphite surface (Figure 5.3f). The passage of current through the graphite causes heating, which is recorded as a temperature. The temperature of the graphite surface can be controlled to preferentially allow destruction and removal of the sample matrix (ashing). Obviously, this may not always be possible and so the addition of matrix modifiers may be required (see Section 4.8.2) to prevent analyte loss. After ashing, the current is rapidly increased to allow vaporisation of the analyte directly into the plasma source. An additional heating stage may be required for cleaning, i.e. the removal of residual material from the graphite surface, prior to cooling. A typical heating process, from sample injection to the cooling stage, may last up to 60 s.

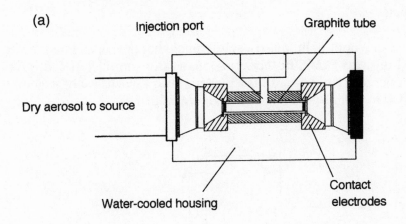

(a)

Injection port

Graphite tube

Dry aerosol to source

Water-cooled housing

Contact electrodes

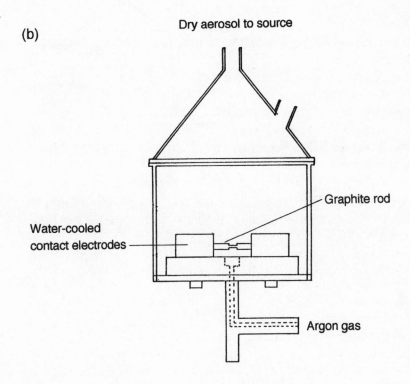

(b)

Dry aerosol to source

Graphite rod

Water-cooled contact electrodes

Argon gas

Figure 5.3f Schematic representation of electrothermal vaporisation: (a) graphite-tube type; (b) graphite-rod type

Laser Ablation

The use of a laser to mobilise a solid sample has been evaluated over a period of many years. In this situation, a laser (normally a Nd–YAG type, operating at 1064 nm), is directed on to the surface of a sample enclosed in a silica-windowed cell (Figure 5.3g).

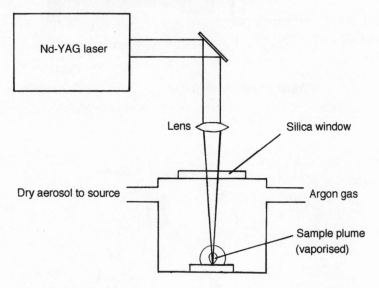

Figure 5.3g   Schematic representation of laser ablation

SAQ 5.3b   The laser most commonly used for ablation is the Nd–YAG type. What does Nd–YAG stand for?

The typical size of the laser crater which is generated is 10–100 pm in diameter. The mobilised sample is transported away from the site of ejection by the argon carrier gas, directly to the plasma source. Depending on the nature of the sample and the power of the laser that is used, the ejection of material may be less than quantitative.

∏ Describe and draw the crater profiles expected after ablation of a hard, e.g. stainless steel, and a soft, e.g. a semiconductor, material.

The hard material's crater profile represents that of a classical volcanic crater with upraised walls, while the softer material will undergo melting and subsequent flowing away from the hot central area.

Sample losses can occur in two main ways: (i) 'hot' sample ejected from the surface can be cooled by the argon carrier gas, thus leading to redeposition; (ii) the ejected material can be lost *en route* to the plasma source by deposition in the connecting tubing. A major disadvantage of laser ablation is the difficulty in obtaining samples for instrument calibration that are suitably homogenous. The limited availability of certified reference materials does not assist the situation, although there are certain cases where suitable samples are available, e.g. steels and alloys.

∏ Suggest a method for producing 'in-house' standards from powdered samples.

For specific purposes, 'in-house' reference samples can be produced by bricketing, i.e. the compression of homogenous powdered samples by using a hydraulic press.

For non-quantitative work, the limited sampling size can be very useful, e.g. for characterising impurities in manufactured goods or for mineral identification in geology.

Flow Injection

Both flow injection (FI) and chromatography (see below) are suitable methods for introducing aqueous samples in a flowing stream into the plasma source or flame, (in AAS). In each case, however, the use of a nebuliser/spray chamber is required for aerosol generation.

Flow injection is a multi-method system that allows the user to create unique sample presentation facilities for the plasma source. In principle, all FI systems consist of a peristaltic pump, injection valve, 'sample alteration/modification unit', and an interface to the nebuliser. The mode of 'sample alteration/modification' is limited only by the ingenuity of the analyst.

SAQ 5.3c	Design a flow injection system for the on-line mixing of two carrier streams. Can you think of a possible way of calibrating a system by using such methodology?

Typically, however, the 'sample alteration/modification' module may consist of one of the following: (i) a low-pressure chromatography column which is used for retention of the analyte in preference to the sample matrix (Figure 5.3h(b)); (ii) a gas–liquid interface for separation of a hydride-generated species (Figure 5.3h(a)) (see Section 5.3.4), a situation in which no nebuliser/spray chamber arrangement will be required; (iii) an on-line solvent extraction system (see Section 2.1.2); (iv) an on-line co-precipitation method (see Section 2.1.3); (v) simply a means of delivering a small discrete sample with the minimum of dilution to the nebuliser/spray chamber.

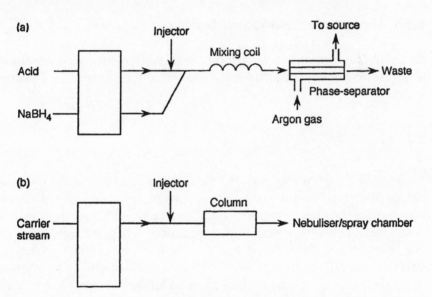

Figure 5.3h Schematic representation of flow injection: (a) hydride generation; (b) chromatography column

Chromatography

The application of chromatography can be useful for two reasons: (i) elemental species information may be obtained, i.e. so-called speciation studies; (ii) potential matrix interferences can be separated. The type of chromatography used largely depends on the nature of the analyte to be separated, but is broadly based on high performance liquid chromatography (HPLC), gas chromatography (GC) and supercritical fluid chromatography (SFC). Within these specific areas, certain variations may be used, e.g. HPLC can include ion-exchange, reverse phase and size exclusion chromatography, while SFC could involve both capillary and packed-column types.

∏   Can you identify what the different types of chromatographic systems could be used for? If you have problems with this, it is recommended that you refer back to the appropriate ACOL texts on *Gas Chromatography* and *High Performance Liquid Chromatography*.

5.3.4 Hydride and Cold-vapour Techniques

Both of these methods of sample introduction are used in atomic absorption spectroscopy and have been described earlier (see Section 4.4.3).

5.4 SPECTROMETERS

Light emitted from the plasma source is focused on to the entrance slit of a spectrometer by using a convex lens arrangement. The function of the spectrometer is to separate the emitted light into its component wavelengths. In practice, depending on the requirements of the analyst and the capital cost of the instrument, two options are available. The first involves a capability to measure one wavelength, corresponding to one element at a time, while the second allows multi-wavelength or multi-element detection. The former is called sequential analysis, or sequential multi-element analysis if the system is to be used to measure several wavelengths one at a time, while the latter is termed simultaneous multi-element analysis. The typical wavelength coverage of a spectrometer for atomic emission spectroscopy is between 167 nm (aluminium) and 852 nm (caesium).

SAQ 5.4a	Would you envisage any difficulties in operating a spectrometer below 190 nm?

SAQ 5.4a

Separation of the light into its component wavelengths is achieved in all modern instruments by the use of a diffraction grating. This consists of a series of closely spaced lines which are ruled or etched on to the surface of a mirror. Most gratings for AES have a line (or groove) density between 600 and 3200 lines per mm. When light strikes the grating it is diffracted at a certain angle, according to the following relationship (the grating equation):

$$n\lambda = d \sin \phi \qquad (5.1)$$

where n is the spectral order, λ is the wavelength, d is the distance between a line (or groove) on the grating, and ϕ is the angle of a groove.

By using equation (5.1) it is possible to calculate the expected wavelength of a spectral line.

SAQ 5.4b   Calculate the wavelength of a spectral emission line, of the first order, with a groove density of 1200 lines per mm and an angle (ϕ) of 20°.

SAQ 5.4b

Interference or ghost images, resulting from overlapping wavelengths, can occur. In order to prevent such spectral overlap, it is possible to use blazed reflection gratings. In this case, the grooves are ruled at a specified angle (known as the blaze angle), and appear as a sawtooth pattern. Thus it is possible to have a blazed diffraction grating which is more efficient for a specific wavelength region. The typical arrangement of a (blazed) diffraction grating is shown in Figure 5.4a; an alternative to this is the Echelle grating, which is described in Section 5.4.2. The resolution of the grating (R) is related to the spectral order (n) and the total number of grooves (N) as follows:

$$R = nN \qquad (5.2)$$

while the resolving power, $\Delta\lambda$, is defined as the wavelength divided by the resolution.

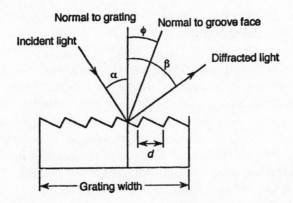

Figure 5.4a Schematic diagram of a blazed grating: d = distance between grooves; ϕ = angle of a groove (blaze angle); α = angle of incidence; β = angle of reflection

As an example, a conventional grating, ruled with 1200 lines per mm and with a width of 52 mm, has a resolution, of the first order, of:

$$R = 1 \times (1200\,\text{mm}^{-1} \times 52\,\text{mm})$$
$$= 62\,400$$

Notice that R has no units.

At a wavelength of 300 nm, the smallest wavelength that can be resolved ($\Delta\lambda$) will be:

$$\Delta\lambda = 300\,\text{nm}/62\,400$$
$$= 0.00481\,\text{nm}$$

5.4.1 Sequential Analysis

A sequential spectrometer is the lower-cost option for atomic emission spectroscopy. This spectrometer typically consists of entrance and exit optics, a diffraction grating and a single detector. A sequential spectrometer has the advantage of flexibility in terms of wavelength coverage. Selection of the desired wavelength is achieved by rotation of the grating within its spectrometer mounting; this operation can be carried out manually or, as is more usual with modern instruments, by computer control.

Π Can you anticipate any potential disadvantages of operating with a sequential spectrometer?

In order to improve the precision of the data which is being collected it is often necessary to include an internal standard (an element that is not present in the sample); other recorded signals can then be referenced to this standard. This reduces any potential interferences from, for example, a change in viscosity from sample to sample. However, in sequential analysis only one element at a time can be monitored, and therefore the use of an internal standard is not possible. (In practice, the use of suitable software will enable two wavelengths to be monitored, thus allowing pseudo-wavelength information to be obtained.)

The Czerny–Turner Mounting

This is the most common spectral mounting for sequential atomic emission spectroscopy. The optical layout of the spectrometer which incorporates this mounting is shown in Figure 5.4b.

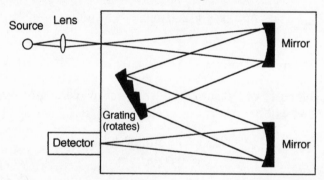

Figure 5.4b Optical layout of the Czerny–Turner spectrometer

5.4.2 Simultaneous Analysis

One of the major advantages of AES is the ability to perform simultaneous multi-element analysis. While this is undoubtedly the higher-cost option it does allow many wavelengths or elements (typically 20–70) to be monitored at the same time. More than one wavelength may be specified for each element if, for example, spectral interference from another element is known to occur. The limitation of such a system is that the exit slits are pre-set, and this fact allows no flexibility if another element and/or wavelength is needed to be analysed.

The Paschen–Runge Mounting

This mounting is the most commonly used polychromator. The grating, entrance slit and multiple exit slits are fixed around the periphery of what is called a Rowland Circle. The grating is concave in appearance and does not rotate. The optical layout of the spectrometer which incorporates these features is shown in Figure 5.4c.

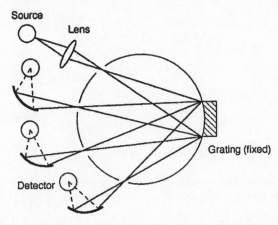

Figure 5.4c   Optical layout of the Paschen–Runge spectrometer

The Echelle Spectrometer

The Echelle spectrometer was originally only commercially available for the direct current plasma (DCP) source. However, with developments in instrumentation the special spectral features that are inherent in its design now provide the analyst with a new generation of compact spectrometer with multi-element capability.

The major component difference of the Echelle spectrometer is the grating. In contrast to the blazed grating described above, this design of grating utilises the spectral order (recall the grating equation) for maximum wavelength coverage. A typical Echelle grating is ruled with ca. 300 lines (or grooves) per mm. The resolution of a diffraction grating is directly related to the groove density (N) and the spectral order (n)

($R = nN$). In this case, however, instead of using a grating with a large number of grooves, resolution is improved by increasing both the blaze angle and the spectral order. A schematic representation of an Echelle grating is shown in Figure 5.4d. Immediately you can see (compare with Figure 5.4a) that the light is reflected off the 'short-side' of the grating. Therefore, the blaze angle is greater than 45°. The advantage of using this method to improve spectral resolution can be seen in Table 5.4, where the various spectral features of both the conventional (blazed) and Echelle gratings are listed.

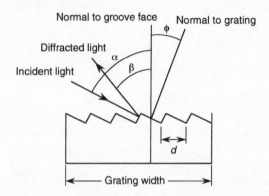

Figure 5.4d Schematic diagram of an Echelle grating: d = distance between grooves, ϕ = angle of a groove (blaze angle); α = angle of incidence; β = angle of reflection

Table 5.4 Comparison of the spectral features of a conventional diffraction grating and an Echelle grating

Feature	Conventional grating	Echelle grating
Focal length (m)	0.5	0.5
Groove or line density (lines per mm)	1200	79
Diffraction angle	10° 22'	63° 26'
Width (mm)	52	128
Spectral order[a]	1	75
Resolution	62 400	758 400
Resolving power[a] (nm)	0.00 481	0.000 396

[a]At 300 nm

However, to prevent any overlapping of the spectral orders a secondary dispersion stage is required. This is typically carried out by using a prism. If the prism is placed so that light separation occurs perpendicular to the diffraction grating (Figure 5.4e), a two-dimensional spectral 'map' is produced where the data is sorted into spectral order vertically and wavelength horizontally (Figure 5.4f).

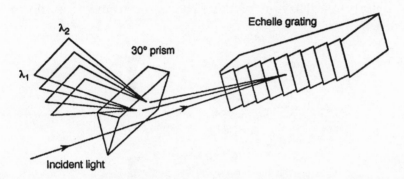

Figure 5.4e Two-dimensional dispersion using a prism in conjunction with an Echelle grating

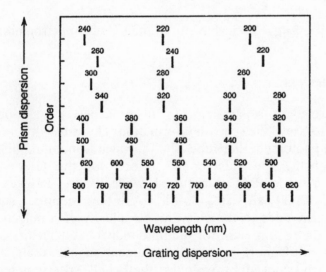

Figure 5.4f Echelle spectrometer: spectral map produced by the arrangement shown in Figure 5.4e

∏   Do you know what type of detection system would allow you to
    observe such a two-dimensional spectral 'map'?

Charge-transfer devices can be used to monitor the detailed
information required to generate such a 'map'; these will be described
in Section 5.4.3 below.

The typical optical layout of the Echelle spectrometer is shown in
Figure 5.4g.

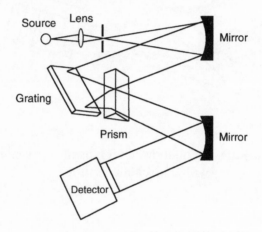

Figure 5.4g   Optical layout of the Echelle spectrometer

5.4.3   Detectors

After wavelength separation has been achieved it is obviously
necessary to 'view' the spectral information. This is usually carried out
by using a photomultiplier tube (PMT), which is mounted behind the
exit slit of the spectrometer. Therefore, for the Czerny–Turner
mounting a single PMT is required, while for the Paschen–Runge
mounting with 30 exists slits, 30 PMTs are now required. This latter
case obviously adds to the cost and complexity of a polychromator
system. If the requirement for multiple individual detectors could be
avoided, a much more powerful instrument should result. This has
recently been achieved; by exploiting the Echelle spectrometer and its
capability to generate a two-dimensional spectral 'map', in
combination with a sensitive multi-wavelength detector, a complete

'fingerprint' of a sample can now be obtained. Several different forms of 'detectors' are available, ranging from the use of photographic film for essentially qualitative analysis, photodiode arrays, the use of which in AES is somewhat restricted due to a lack of sensitivity, through to the new generation of charge-transfer devices.

Photomultiplier Tube

This is a device that converts incident light into a current. This is carried out by a series of steps which are outlined in Figure 5.4h.

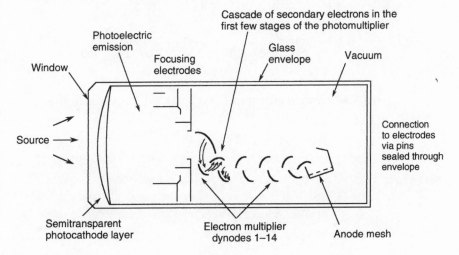

Figure 5.4h   Schematic representation of the operation of a photo-multiplier tube

Incident light first passes through the silica window and strikes the photocathode. The latter consists of an easily ionised material such as an alloy of two (or three) alkali metals with antimony. The exact composition of the photocathode affects the wavelength range of a particular detector. For example, in Figure 5.4i it can be seen that the use of bialkali (Sb–K–Cs) provides a greater wavelength coverage than does the use of high-temperature bialkali (Na–K–Sb). In this way, it is possible to select a PMT that has optimum response characteristics for a particular wavelength range. Such a feature can easily be exploited in a polychromator where an individual PMT is used to monitor only one wavelength.

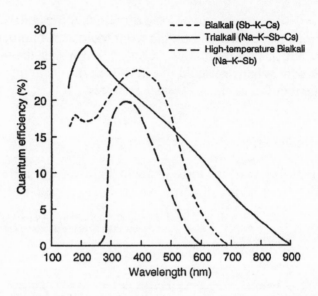

Figure 5.4i   Spectral responses of selected photocathodes

SAQ 5.4c   After considering the spectral information given in Figure 5.4i, which do you think would be the best photomultiplier tube for use in a monochromator?

A suitably energetic incident photon of light causes an electron to be emitted from the cathode surface. This process is known as the photoelectric effect. The yield per incident photon is called the quantum efficiency, with typical values lying between 10 and 25% for the range 650–340 nm. The emitted electron is then directed towards the dynode chain by a series of focusing electrodes. The action of the dynode chain is to multiply this single electron into many electrons. This is achieved as follows. A single electron striking dynode 1 will emit at least two secondary electrons. These electrons will then strike dynode 2, with each electron that strikes this second dynode causing at least a further two electrons to be emitted, and so on. In this way, a cascade of electrons is produced (see Figure 5.4h). The exact number of electrons that are generated will depend on the length of the dynode chain, which typically consists of 9–16 dynodes. This amplification of electrons by the dynode chain is known as the 'gain'; a typical gain for a nine-dynode PMT is 10^6. The final step is to collect the electrons at the anode; the electrical current that is now measured at the anode is proportional to the amount of light that struck the photocathode. This current is then converted into a voltage signal which is then transferred via an analog-to-digital (A/D) converter to a suitable computer for processing purposes.

Charge-transfer Devices

The utilisation of so-called charge-transfer devices (CTDs) is perhaps the most significant advance in detector technology for atomic emission spectroscopy.

∏ Have you come across the use of a charge-transfer device while on your holidays?

It is quite likely that you have, as camcorders use a charge-coupled device (CCD), which is a form of CTD, to view the image.

Charge-transfer devices offer high sensitivity and a wide wavelength coverage. Their main application has been as detectors for the Echelle spectrometer. The compact nature of the spectral 'map' generated by the Echelle spectrometer can be focused on to a single CTD. Essentially, a CTD consists of an array of closely spaced metal–insulator–semiconductor diodes formed on a wafer of semiconductor material. Two common forms are available, namely the charge-coupled device (CCD) and the charge-injection device (CID). In both types, incident light is converted into a signal. In essence, the CTD acts as an electronic photographic film. A full description of the mode of operation of CTDs and the particular characteristics of CIDs and CCDs is unfortunately beyond the scope of this present text.

5.5 INTERFERENCES

The spectral interferences which occur in atomic emission spectroscopy can be classified into two main categories, i.e. spectral overlap and matrix effects. Spectral interferences are probably the most well known and best understood. The usual remedy to alleviate a spectral interference is to either increase the resolution of the spectrometer, or to select an alternative spectral emission line. Three types of spectral overlap can be identified (Figure 5.5):

(a) direct wavelength coincidence with an interfering emission line;

(b) partial overlapping of the emission line of interest with an interfering line in close proximity;

(c) the presence of an elevated or depressed background continuum.

Type (a) and (b) interferences can occur as a result of an interfering emission line from another element, the source gas (Ar) or impurities within or entrained in the source, e.g. molecular species such as OH, N_2, etc. Extensive work by various groups has meant that wavelength coincidence (type (a)) is well characterised for the ICP source. Examples of this are Cd (228.802 nm) and As (228.812 nm) and Zn (213.856 nm) and Ni (213.858 nm). Elimination of type (b) is usually only possible by an improvement in resolution. As this may not be possible on a routine basis, mathematical models can be used to try and correct for this type of interference. The only certain remedy, however, is to select an interference-free wavelength for the selected element. Type (c) can be corrected for by measurement of the background on either side of the wavelength of interest. Provided that no significant fine structure is present on the background, this method of correction should prove to be satisfactory.

Matrix interferences are often associated with the sample introduction process. For example, pneumatic nebulisation can be affected by the

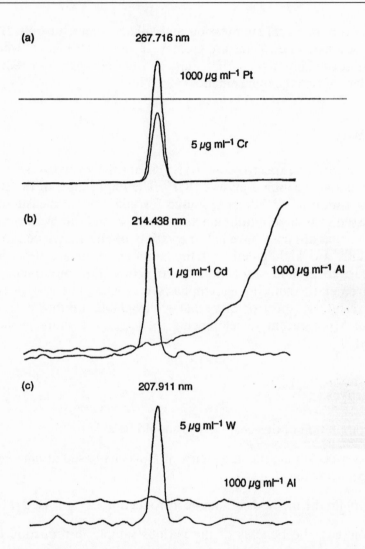

Figure 5.5 Spectral interferences: (a) spectral overlap; (b) wing overlap; (c) background shift

dissolved-solids content of the aqueous sample, which can then affect the uptake rate of the nebuliser, and hence the sensitivity of the determination. Matrix effects that are encountered in the plasma source have also been documented. Typically, this involved the presence of easily ionisable elements (EIEs), e.g. alkali metals, within the plasma source. Specific work has investigated the effect of these

EIEs on both signal suppression and enhancement for both ICP and DCP sources. The effects are greatest in the latter source, where the addition of lithium or barium salts as buffers is used to reduce the problem of signal enhancement.

Summary

Atomic emission spectroscopy (AES) is normally carried out by using an inductively coupled plasma (ICP). The importance of the ICP and its utilisation as an effective source for quantitative measurement is discussed. The high requirements needed for the sample introduction devices that are used have led to a variety of alternative designs. Such demands are highlighted and the methods used are described. In contrast to atomic absorption spectroscopy, the importance of the spectrometer cannot be overemphasised in AES. The various types of high-resolution spectrometers that are available are thus described in detail. More recent developments in detector technology are also included.

Objectives

On completion of this chapter you should be able to:

● describe the main components of a plasma-based atomic emission spectrometer;

● explain the methods of sample introduction devices used for AES;

● compare the benefits of the various sample introduction devices that are used;

● identify why it is necessary to use a spray chamber;

● compare continuous and discrete sample introduction techniques;

● explain why the microwave-induced plasma is used as a detector for gas chromatography;

● describe the merits of sequential and simultaneous multi-element detection;

- identify the benefits of using an Echelle spectrometer;

- draw the optical arrangements of selected spectrometers;

- calculate related diffraction grating parameters, including resolution;

- describe the operation of a photomultiplier tube;

- suggest reasons why charge-transfer devices are useful in AES;

- identify interferences and their remedies in AES.

6. Inorganic Mass Spectrometry

6.1 INTRODUCTION

Within the scope of this present book the traditional techniques of spark source and thermal ionisation mass spectrometry are not covered. The interested reader is therefore referred to more specialist texts for relevant information.

6.2 INDUCTIVELY COUPLED PLASMA MASS SPECTROMETRY

In contrast to other sources used for inorganic mass spectrometry, the inductively coupled plasma offers several advantages, not least the ability to analyse samples rapidly. This major advantage, coupled with the high degree of sensitivity offered by inductively coupled plasma mass spectrometry (ICP-MS), are the unique features which have allowed this technique to develop from initial conception, to commercial development and then through to its routine application in a relatively short time-scale (Figure 6.2a).

6.2.1 Principle of Operation

The major instrumental development which was required in order to fully establish ICP-MS as a major analytical technique was the efficient coupling of an ICP, which operates at atmospheric pressure, with a mass spectrometer, which operates under high vacuum (Figure 6.2b). The development of a suitable interface held the key to the establishment of this technique.

An additional feature of the mass spectrometer is the ability to measure isotope ratios. The importance of this feature is readily

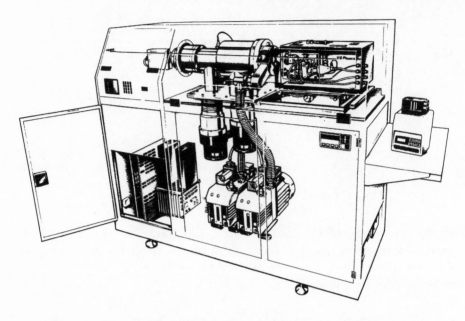

Figure 6.2a Inductively coupled plasma mass spectrometer

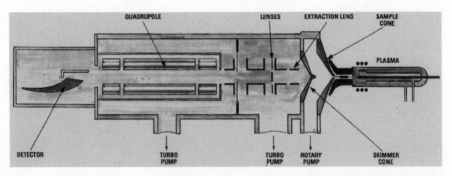

Figure 6.2b Schematic diagram of an inductively coupled plasma mass spectrometer

appreciated when you consider that ca. 70% of the elements in the periodic table have stable (i.e. non-radioactive) isotopes. The ability to measure isotope ratios has two major benefits. The first is that it allows stable isotopes with an altered isotopic ratio to be used for tracer studies, e.g. for monitoring the absorption of nutrients in the human body without the use of radio-labelled isotopes. Secondly, the use of enriched stable isotopes allows calibration of an unknown sample without the requirement to prepare solutions for direct calibration or standard additions. The method used here is known as isotope dilution analysis (see Section 6.4).

6.2.2   Ion Source: ICP

The ion source used in this technique is the inductively coupled plasma (ICP). Its formation and operation have been described earlier (see Section 5.2.1). Indeed, it is the same source that is used for atomic emission spectroscopy.

∏   What is the difference between the ICP torch configuration used in mass spectrometry and that used in atomic emission spectroscopy (see Figure 5.2a in Section 5.2.1)?

The only instrumental modification is in how the analyte is observed. In AES, the torch is positioned vertically, so that emission is (normally) observed at right angles by the spectrometer. In MS, the torch is positioned horizontally so that ions can be extracted from the ICP directly into the spectrometer (Figure 6.2c). As a consequence of this horizontal positioning of the ICP torch in relation to the mass spectrometer, all species that enter the plasma are transferred into the machine. In addition, all of the sample introduction devices used for ICP-AES can be similarly used for ICP-MS.

∏   Recall the sample introduction devices used for ICP-AES. Which is the most popular device?

In ICP-AES the most common method for introducing the sample is via a pneumatic nebuliser/spray chamber assembly. This is also the preferred method for ICP-MS.

6.2.3   Interface

The main reason for in the success of the ICP-MS technique has been the development of a suitable interface.

Figure 6.2c The ICP torch box and sampling cone of a mass
spectrometer

∏ What would you forsee as the main problem in interfacing an ICP
with a mass spectrometer?

The interface allows the coupling of the ICP source (at atmospheric
pressure) with the mass spectrometer (at high vacuum), while still
maintaining a high degree of sensitivity. The interface consists of a
water-cooled outer sampling cone which is positioned in close
proximity to the plasma source (Figure 6.2d). The sampling cone is
made of nickel because of its high thermal conductivity, relative
resistance to corrosion, and robust nature. The pressure differential
created by the sampling cone is such that ions from the plasma and the
plasma gas itself are drawn into the region of lower pressure through
the small orifice (1.0 mm) of the cone. The region behind the sampling
cone is maintained at a moderate pressure (ca. 2.5 mbar (250 Pa)) by
using a rotary vacuum pump. As the gas flow through the sampling

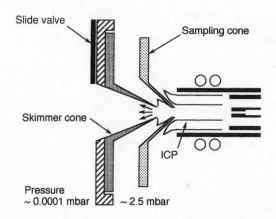

Figure 6.2d Schematic diagram of the inductively coupled plasma/ mass spectrometer interface

cone is large, a second cone is placed close enough behind it to allow the central portion of the expanding jet of plasma gas and ions to pass through the skimmer cone. The latter, which is also made of nickel, has an orifice diameter of ca. 0.75 mm. The pressure behind the skimmer cone is maintained at ca. 10^{-4} mbar (10^{-2} Pa). The extracted ions are then focused by a series of electrostatic lenses into the mass spectrometer.

6.2.4 Mass Spectrometer

The mass spectrometer acts as a filter, transmitting ions with a preselected mass/charge ratio. These transmitted ions are then detected with a channel electron multiplier.

SAQ 6.2a What type of mass spectrometer do you think could be used in ICP-MS?

SAQ 6.2a

Quadrupole Instrument

The quadrupole analyser consists of four straight metal rods positioned parallel to and equidistant from a central axis (Figure 6.2e). By applying direct current (DC) and radio frequency (RF) voltages to opposite pairs of the rods it is possible to have a situation where the DC voltage is positive for one pair and negative for the other. In a similar way, the RF voltages on each pair are then 180° out of phase, i.e. they are opposite in sign, but with the same amplitude. Ions entering the quadrupole are subjected to oscillatory paths by the RF voltage. However, by selecting the appropriate RF and DC voltages only ions of a given mass/charge ratio will be able to traverse the length of the rods and emerge at the other end. Other ions are lost within the quadrupole analyser; if their oscillatory paths are too large they will collide with the rods and become neutralised.

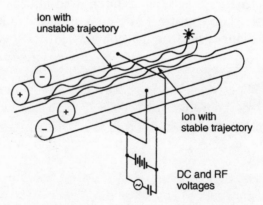

Figure 6.2e   Schematic diagram of the quadrupole analyser arrangement

SAQ 6.2b | Where do the neutralised ions end up in the system?

ICP-MS can be carried out in two distinctly different modes, i.e. with the mass filter transmitting only one mass/charge ratio, or with the DC and RF voltages being changed continuously. The former case would allow single-ion monitoring, with the latter allowing multi-element analysis. In single-ion monitoring, all of the data are obtained from single mass/charge ratio measurements; although this precludes the major facet of the technique it does provide a higher degree of sensitivity for the element (mass/charge) of interest. For multi-element analysis, RF and DC voltage scanning is required. Scanning, and hence data acquisition, can be carried out in several modes. The possibilities are as follows: (a) a single continuous scan; (b) peak hopping; (c) multi-channel scanning. These different scanning modes are illustrated, using silver as an example (^{107}Ag is 51.8% abundant and ^{109}Ag is 48.2% abundant), in Figure 6.2f. In single continuous scanning, the mass/charge ratio is changed continuously in one scan. However, in order to reduce ion fluctuations and improve precision it is better if the mass/charge ratio is scanned repeatedly. This can be done by using either peak hopping or by multi-channel scanning. In peak hopping, the signal ions are measured at selected mass/charge ratios for a particular dwell time (0.5–1 s). This allows the analyst to carry out fast repetitive analyses of a predetermined set of elements. However, it does not allow any interrogation of the mass spectrum for potential interferences, e.g. unexpected polyatomic interferences (see Section 6.2.6). In multi-channel scanning, all mass/charge ratios are repeatedly scanned, thus providing a complete 'fingerprint' of the

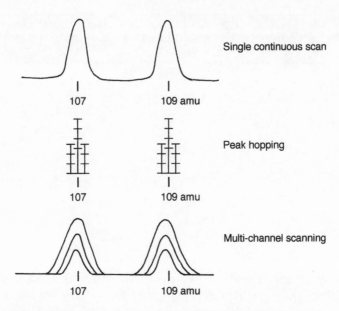

Figure 6.2f Methods of data acquisition in inductively coupled
plasma mass spectrometry

unknown sample composition. This scanning mode can also be useful
for the quick qualitative analysis of unknown samples. In the multi-
channel scanning mode, the typical dwell time per mass/charge ratio is
0.1–0.5 ms. The quadrupole mass analyser is therefore a very rapid
sequential spectrometer.

SAQ 6.2c   What is the difference between sequential and
simultaneous multi-element analysis? What analogy
does this provide in ICP optical spectroscopy?

SAQ 6.2c

Quadrupole mass analysers are capable of only unit mass resolution, i.e. they can only measure integral values of the mass/charge ratio (e.g. 204, 205, 206, etc.). An important criterion in ICP-MS is the ability to separate a weak signal intensity at mass M from an adjacent major peak, i.e. at $M+1$ or $M-1$. This is termed abundance sensitivity. As you might imagine, this is an important factor in ICP-MS when ultratrace analysis is being carried out against a background of a major element impurity (or the sample matrix) — the proverbial needle (ultratrace element of interest) in a haystack (sample matrix). Values of up to 10^6 to 1 are achievable with quadrupole ICP-MS instruments.

Double-focusing High-resolution Instrument

The lack of resolution for a quadrupole instrument (i.e. unit mass resolution) makes it impossible to separate interferences (see Section 6.2.6) that coincide at the same nominal mass. A spectral resolution of ca. 10 000 would remove most of the interferences from polyatomic species, but much higher resolution is required for isobaric interferences (Table 6.2a). An instrument which is capable of separating at such high resolutions is the double-focusing magnetic sector mass spectrometer (Figure 6.2g). However, the high cost of these instruments has precluded their wide acceptance.

Table 6.2a Some examples of the resolution required to separate ions of similar intensity[a]

Analyte ion	Interfering ion	Resolution[b]
$^{28}Si^+$	$^{14}N_2^+$	960
$^{56}Fe^+$	$^{40}Ar^{16}O^+$	2 500
$^{80}Se^+$	$^{40}Ar_2^+$	9 700
$^{58}Ni^+$	$^{58}Fe^+$	28 000
$^{87}Rb^+$	$^{87}Sr^+$	300 000

[a]Data taken from K.E. Jarvis, A.L. Gray and R.S. Houk, *The Handbook of Inductively Coupled Plasma Mass Spectrometry*, Blackie Academic and Professional, Glasgow, 1992, p. 46
[b]Resolution = $M/\Delta M$

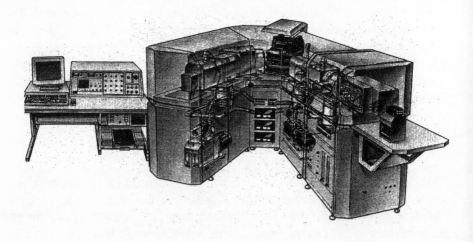

Figure 6.2g High-resolution inductively coupled plasma (double-focusing magnetic sector) mass spectrometer

6.2.5 Detector

The most common type of detector used in quadrupole instruments is the channeltron electron multiplier (Figure 6.2h). The operating principles of the electron multiplier are similar to those of the photomultiplier tube (see Section 5.4.3), apart from the absence of dynodes. In addition, the electron multiplier must operate under vacuum conditions ($< 5 \times 10^{-5}$ torr). The device consists of an open tube with a wide cone entrance. The inside of the tube is coated with lead oxide (a semiconducting material). The cone is biased with a high negative potential (ca. $-3\,\text{kV}$) at the entrance and held at ground near the collector. Any incoming positive ion, from the mass analyser, is attracted towards the negative potential of the cone. On impact, the positive ion causes one or more (secondary) electrons to be ejected, which are then attracted towards the grounded collector within the tube. In addition, the initial secondary electrons can also collide with the surface coating, thus causing further electrons to be ejected. This multiplication process continues until all of the electrons (up to 10^8) are collected. This discrete pulse of electrons is further amplified exterior to the electron multiplier tube and recorded as the number of ion counts per second. All electron multiplier tubes have a limited lifetime, which is determined by the total accumulated charge which is monitored.

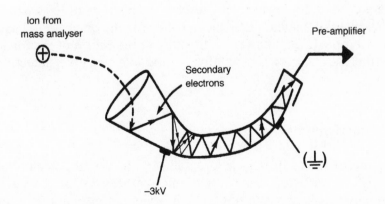

Figure 6.2h Operating principles of the electron multiplier tube

SAQ 6.2d   Compare the operation of the electron multiplier tube with that of the photomultiplier tube.

The electron multiplier can also respond to photons of light from the ICP. In order to alleviate this, the detector (or the quadrupole) can be mounted off-axis, or a baffle can be fitted in the centre of the ion lens. Nevertheless, the electron multiplier tube is a sensitive detector for ICP-MS.

6.2.6 Interferences

While interferences in ICP-quadrupole MS are not so prevalent as in ICP-AES, nevertheless some type of interferences do occur. These can be broadly classified, according to their origin, into isobaric, molecular and matrix dependent. As in AES, and for that matter AAS, some overlap of elements can occur (isobaric), and again as in these other techniques the relevant data are well documented. However, other types of interference can occur in ICP-MS; these result from the acid(s) used to prepare the sample and/or the argon plasma gas (polyatomics). In addition, the formation of oxide, hydroxide and doubly charged species is possible. Finally, the occurrence of matrix interferences can result in signal enhancement or depression with respect to the atomic mass. Each of these types of interferences will be discussed below.

Isobaric Interferences

This type of interference is well characterised (Table 6.2b), and as a result of the fact that ca. 70% of the elements in the periodic table have more than one isotope, it can usually be avoided by selecting an alternative isotope. By examining Table 6.2b it is possible to identify situations in which potential problems can be alleviated by the ability to select an alternative isotope.

Table 6.2b Isobaric interferences for elements from period 4 of the periodic table

Atomic mass	Element of interest[a]	Interfering element[a]
39	K (93.10)	–
40	Ca (96.97)	Ar (99.6)[b]; K (0.01)
41	K (6.88)	–
42	Ca (0.64)	–
43	Ca (0.14)	–
44	Ca (2.06)	–
45	Sc (100)	–
46	–	Ca (0.003); Ti (7.93)
47	–	Ti (7.28)
48	Ti (73.94)	Ca (0.19)
49	–	Ti (5.51)
50	–	Ti (5.34); V (0.24); Cr (4.31)
51	V (99.76)	–
52	Cr (83.76)	–
53	–	Cr (9.55)
54	–	Cr (2.38); Fe (5.82)

continued overleaf

Table 6.2b (*continued*)

Atomic mass	Element of interest[a]	Interfering element[a]
55	Mn (100)	–
56	Fe (91.66)	–
57	–	Fe (2.19)
58	Ni (67.88)	Fe (0.33)
59	Co (100)	–
60	Ni (26.23)	–
61	–	Ni (1.19)
62	–	Ni (3.66)
63	Cu (69.09)	–
64	Zn (48.89)	Ni (1.08)
65	Cu (30.91)	–
66	Zn (27.81)	–
67	–	Zn (4.11)
68	–	Zn (18.57)
69	Ga (60.40)	–
70	–	Zn (0.62); Ge (20.52)
71	Ga (39.60)	–
72	Ge (27.43)	–
73	–	Ge (7.76)
74	Ge (36.54)	Se (0.87)
75	As (100)	–
76	–	Ge (7.76); Se (9.02)
77	–	Se (7.58)
78	Se (23.52)	Kr (0.35)
79	Br (50.54)	–
80	Se (49.82)	Kr (2.27)
81	Br (49.46)	–
82	–	Se (9.19); Kr (11.56)
83	Kr (11.55)	–
84	Kr (56.90)	Sr (0.56)[c]
85	–	–
86	–	Kr (17.37); Sr (9.86)[c]

[a]Percentage abundance is shown in parentheses
[b]Not in the 4th period of the periodic table but included because of its origin in the plasma source
[c]Not in the 4th period of the periodic table, but included for the sake of completeness

For example, if you wanted to analyse nickel in a stainless steel sample, it would seem appropriate to select the most abundant isotope for this element. This occurs at atomic mass 58, where nickel is 67.8% abundant. However, iron (0.31% abundant) also occurs at this same mass. Therefore, in order to prevent any isobaric interference it is necessary to select an alternative mass. At atomic mass 60, nickel is 26.4% abundant and no interfering isotopes now occur. One other point to consider is that the selection of a less abundant isotope for nickel has potentially led to a loss in sensitivity. This latter point may not be significant if nickel is present at a high enough concentration in the steel sample because of the inherent sensitivity of the technique.

SAQ 6.2e Can you identify a situation, in which an alternative mass could be used, for the determination of zinc in a nickel alloy?

SAQ 6.2f Can you also identify a similar situation for titanium, in which the only available alternative mass is one with an inherently low percentage abundance?

SAQ 6.2f

Alternatively, other situations can exist which do not have such an easy solution. Probably the best example of this is the determination of calcium (atomic mass 40). Unfortunately (for this element), the ICP source consists of argon ions which have a mass coincidence at atomic mass 40 (Ar, 99.6% abundant). In this particular case, no alternative mass exists for Ca which will provide any degree of sensitivity. The best available mass is 44, at which Ca has an abundance of 2.08%.

SAQ 6.2g   Can you identify a situation for selenium in which no alternative mass is available?

Molecular Interferences

There are two main types of molecular interferences, namely those derived from the formation of (a) polyatomic moieties and (b) doubly charged species.

Polyatomics. These interferences occur as a result of selected interactions between the element of interest and its associated aqueous solution, the plasma gas itself (Ar) and the type of acid(s) used to digest or prepare the sample (Table 6.2c). The latter includes interferences from interactions with nitric, sulfuric, hydrochloric and phosphoric acids. At this point, it should now be evident that the spectrum obtained from a quadrupole mass spectrometer may be more complicated than originally anticipated. Very few elements in period 4 of the periodic table are unaffected by some type of interference (either isobaric or polyatomic). The true origin of the polyatomic ions detected by the spectrometer, however, is not known, although they are not believed to be associated with the very hot (10 000 K) plasma source. It is more likely, therefore, that they are formed within the interface where the ions are undergoing transfer from the atmospheric source to the vacuum conditions of the spectrometer. Nevertheless, they provide an unwelcome additional aspect to the interpretation of the mass spectra. As a consequence, the unsuspecting analyst may record an enhanced signal which is due to the presence of a polyatomic interference.

Table 6.2c Potential polyatomic interferences

Atomic mass	Element of interest[a]	Polyatomic interference
39	K (93.10)	$^{38}Ar^1H^+$
40	Ca (96.97)	$^{40}Ar^+$
41	K (6.88)	$^{40}Ar^1H^+$
42	Ca (0.64)	$^{40}Ar^2H^+$
43	Ca (0.14)	–
44	Ca (2.06)	$^{12}C^{16}O^{16}O^+$
45	Sc (100)	$^{12}C^{16}O^{16}O^1H^+$
46	–	$^{14}N^{16}O^{16}O^+$; $^{32}S^{14}N^+$
47	–	$^{31}P^{16}O^+$; $^{33}S^{14}N^+$
48	Ti (73.94)	$^{31}P^{16}O^1H^+$; $^{32}S^{16}O^+$; $^{34}S^{14}N^+$
49	–	$^{32}S^{16}O^1H^+$; $^{33}S^{16}O^+$; $^{14}N^{35}Cl^+$
50	–	$^{34}S^{16}O^+$; $^{36}Ar^{14}N^+$
51	V (99.76)	$^{35}Cl^{16}O^+$; $^{34}S^{16}O^1H^+$; $^{14}N^{37}Cl^+$ $^{35}Cl^{16}O^+$
52	Cr (83.76)	$^{40}Ar^{12}C^+$; $^{36}Ar^{16}O^+$; $^{36}S^{16}O^+$; $^{35}Cl^{16}O^1H^+$

(continued overleaf)

Table 6.2c (*continued*)

Atomic mass	Element of interest[a]	Polyatomic interference
53	–	$^{37}Cl^{16}O^+$
54	–	$^{40}Ar^{14}N^+$; $^{37}Cl^{16}O^1H^+$
55	Mn (100)	$^{40}Ar^{14}N^1H^+$
56	Fe (91.66)	$^{40}Ar^{16}O^+$
57	–	$^{40}Ar^{16}O^1H^+$
58	Ni (67.88)	–
59	Co (100)	–
60	Ni (26.23)	–
61	–	–
62	–	–
63	Cu (69.09)	$^{31}P^{16}O_2^+$
64	Zn (48.89)	$^{31}P^{16}O_2^1H^+$; $^{32}S^{16}O^{16}O^+$; $^{32}S^{32}S^+$
65	Cu (30.91)	$^{33}S^{16}O^{16}O^+$; $^{32}S^{33}S^+$
66	Zn (27.81)	$^{34}S^{16}O^{16}O^+$; $^{32}S^{34}S^+$
67	–	$^{35}Cl^{16}O^{16}O^+$
68	–	$^{40}Ar^{14}N^{14}N^+$; $^{36}S^{16}O^{16}O^+$; $^{32}S^{36}S^+$
69	Ga (60.40)	$^{37}Cl^{16}O^{16}O^+$
70	–	$^{35}Cl_2^+$; $^{40}Ar^{14}N^{16}O^+$
71	Ga (39.60)	$^{40}Ar^{31}P^+$; $^{36}Ar^{35}Cl^+$
72	Ge (27.43)	$^{37}Cl^{35}Cl^+$; $^{36}Ar^{36}Ar^+$; $^{40}Ar^{32}S^+$
73	–	$^{40}Ar^{33}S^+$; $^{36}Ar^{37}Cl^+$
74	Ge (36.54)	$^{37}Cl^{37}Cl^+$; $^{36}Ar^{38}Ar^+$; $^{40}Ar^{34}S^+$
75	As (100)	$^{40}Ar^{35}Cl^+$
76	–	$^{40}Ar^{36}Ar^+$; $^{40}Ar^{36}S^+$
77	–	$^{40}Ar^{37}Cl^+$; $^{36}Ar^{40}Ar^1H^+$
78	Se (23.52)	$^{40}Ar^{38}Ar^+$
79	Br (50.54)	$^{40}Ar^{38}Ar^1H^+$
80	Se (49.82)	$^{40}Ar^{40}Ar^+$
81	Br (49.46)	$^{40}Ar^{40}Ar^1H^+$
82	–	$^{40}Ar^{40}Ar^1H^1H^+$
83	Kr (11.55)	–
84	Kr (56.90)	–
85	–	–
86	–	–

[a]Percentage abundance is shown in parentheses

Doubly charged species. A further type of polyatomic interference results from the formation of doubly charged species. Of particular note here are the ionic species of cerium, lanthanum, strontium, thorium and barium.

SAQ 6.2h

> Barium has atomic mass values of 130, 132, 134, 135, 136, 137, and 138. By using this information, can you identify the atomic masses at which Ba^{2+} species will occur?

Matrix-dependent Interferences

Matrix dependency problems usually result from the presence of excess salts (e.g. NaCl) in the ICP. Two effects have been observed, usually in terms of a loss of sensitivity, for specific elements. The first is related to the plasma source region and the second to mass discrimination effects in the spectrometer. In the first case, an incorrect choice of the appropriate sample introduction device can lead to blockage problems in the nebuliser. However, once this problem has been successfully resolved, solids may also build up on the sample cone of the ICP-MS interface. Both of the above can lead to intermittent and erratic signal generation. Nevertheless, remedies can be sought and used, in the former case by choosing the appropriate nebuliser, e.g. a high-solids type. The second problem can be overcome by either aqueous dilution of the sample matrix (which may lead to decreased sensitivity of the analyte of interest), or by using flow injection (the intermittent introduction of a sample with a high salt content, followed by washing in dilute acid). Matrix matching of samples with calibration solutions may also be advantageous. Although mass discrimination effects (which usually result in a lower sensitivity) can be experimentally observed in the mass spectrometer, their exact mechanism is not known. A recent approach to alleviate the problem of the matrix has been the use of tandem techniques. (A tandem technique is the combination of two or more different instruments to produce a hybrid instrument.) In this present case, the combination of a chromatographic separation technique with ICP-MS can offer some advantages. Use of the chromatographic method may allow preferential separation of the matrix from the element of interest by, for example, an ion–exchange process or, if the element has been chemically attached to an organic molecule, by reverse phase high performance liquid chromatography. While the use of tandem techniques implies the direct coupling of one or more different instruments, such procedures could take place in the off-line mode, with separation being carried out remotely from the ICP-MS; normal sample introduction to the ICP is then made at a later stage.

6.3 GLOW DISCHARGE MASS SPECTROMETRY

Unlike the inductively coupled plasma, a glow discharge source acts directly upon a solid sample. The most common glow discharge (GD) source is based on the application of a direct current (DC), although an alternative GD source based on radiofrequency (RF) has been reported in the scientific literature. With the DC glow discharge source, the solid sample must be conducting, e.g. in the analysis of steel. However, if a non-conducting sample is to be analysed, then prior mixing with a conducting substrate, e.g. copper powder, and compaction will be required. Glow discharge sources are used in atomic emission spectroscopy as well as in mass spectrometry.

6.3.1 Ion Source

Glow discharges are essentially low-pressure plasmas that rely on cathodic sputtering to atomise a solid sample. The atomised sample can then undergo collision with electrons and metastable fill-gas atoms, thus causing excitation and ionisation. A schematic diagram illustrating the operation of a glow discharge source is shown in Figure 6.3a.

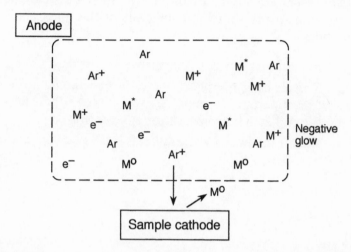

Figure 6.3a Operating principles of a glow discharge source

The flow of electric current through the GD source initiates the analytical process. As the negatively charged electrons are attracted towards the anode (positive terminal), collisions with the fill-gas (argon) atoms can occur, producing argon ions. The coexistence of argon ions, neutral argon atoms and electrons constitutes a plasma (see Section 5.2.1; ICP formation). The initiation of this plasma provides an integral site for further ionisation and excitation. Concurrent with plasma formation, the positively charged argon ions are attracted towards the cathode. As the conducting sample and cathode are one and the same, the argon ions are able to collide with the sample surface, liberating metal atoms, a process known as sputtering (see earlier). The liberated metal sample atoms can then diffuse into the plasma for excitation/ionisation (AES) or ionisation (MS). The two most probable principle ionisation mechanisms are as follows:

$$M + e^- \longrightarrow M^+ + 2e^- \text{ (electron impact)} \tag{6.1}$$

$$M + Ar^* \longrightarrow M^+ + Ar + e^- \text{ (Penning exchange)} \tag{6.2}$$

where M is the sputtered atom and Ar^* is an excited-state (metastable) argon atom.

Penning-exchange ionisation is believed to be the dominant process, and leads to fairly uniform ionisation sensitivity. The generated ions are extracted from the source by the use of an electrostatic field and then focused on to the source-defining slit of the mass analyser by using ion optics.

SAQ 6.3a	How does the operation of a glow discharge source compare with that of a hollow-cathode lamp (as used in atomic absorption spectroscopy)?

The rate of sputtering is dependent on several variables, namely the material to be analysed, the fill gas, the applied current and voltage, and the gas pressure. A schematic diagram of a typical GD source used for mass spectrometry is shown in Figure 6.3b. Typical operating conditions are as follows: current, 4 mA; voltage, 1.2 kV; pressure, 2×10^{-4} mbar (0.02 Pa).

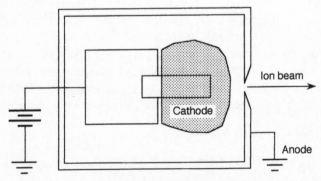

Figure 6.3b Schematic diagram of a typical glow discharge source used for mass spectrometry

SAQ 6.3b	Why should changing from an argon to an helium fill-gas affect sputtering?

6.3.2 Spectrometer

The mass analyser or spectrometer is normally a double-focusing instrument of reverse Nier–Johnson geometry (Figure 6.3c). This instrument has both a magnetic and an electrostatic analyser. Other mass spectrometer types can also be used, e.g. the quadropole, but the poor resolution that is achieved does not preclude the effect of interferences (see Section 6.2.6). It has been reported that potential molecular interferences can be removed to some extent by cryogenic treatment of the source.

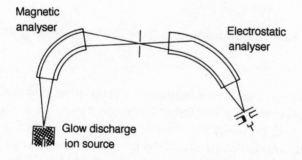

Figure 6.3c Schematic diagram of a high-resolution glow discharge mass spectrometer

6.3.3 Applications

The main applications of GD-MS (Figure 6.4) are principally in the area of surface analysis, although bulk analysis is also possible. By careful control of the operating conditions, it is possible to erode multiple-layer samples while monitoring the elemental composition of the material. In this way, the thickness of layers of different elemental composition can be determined. This is particularly important in laminated substances, e.g. semiconductors.

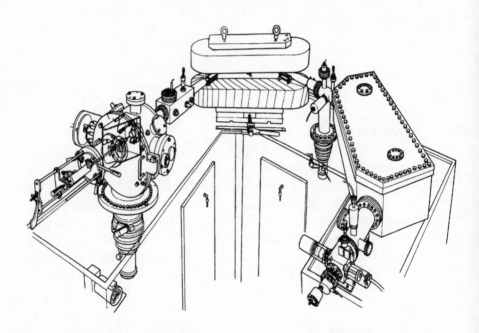

Figure 6.3d   High-resolution glow discharge mass spectrometer

As an example, the determination of silicon in an iron matrix by using a low-resolution glow discharge mass spectrometer would not allow differentiation between the doubly charged iron ($^{56}Fe^{2+}$) species and the gaseous impurities from carbon monoxide and nitrogen present on the silicon (^{28}Si). However, as can be seen in Figure 6.3e, the use of a high-resolution mass spectrometer allows the silicon to be successfully determined.

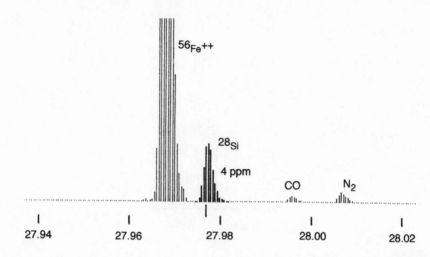

Figure 6.3e Determination of silicon in steel using high-resolution GD-MS

SAQ 6.3c	Explain why the doubly charged ion species, $^{56}Fe^{2+}$, occurs at 28 atomic mass units (amu).

A similar example is given in Figure 6.3f, which shows the separation of the polyatomic interference, ArO^+, from the Fe^+ signal at 56 amu. The use of a high-resolution mass spectrometer allows separation of both species, and thus Fe^+ can be unequivocally determined.

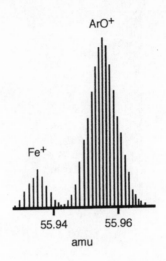

Figure 6.3f   Determination of iron using high-resolution GD-MS

6.4   ISOTOPE DILUTION ANALYSIS

The normal methods of calibration for atomic spectroscopy are external calibration and standard additions (see Section 1.3). However, the use of a mass spectrometer provides an alternative method, namely isotope dilution analysis (IDA). All mass spectrometers are capable of measuring isotope ratios (the ratios of the different isotopes of a single element), and IDA uses this approach to provide a unique method of calibration. The essential feature of the technique of IDA is that the element under investigation has more than one stable isotope; this applies to more than 75% of the elements of the periodic table. The basis of the approach is that the isotope ratios are measured before (i.e. the original sample) and after spiking (sample plus spike), and by mathematical treatment of the results obtained the concentration of the element in the original sample can then be determined. A key factor in this process is the nature of the spike which is added. The spike must have an artificially enriched isotopic abundance of the element of interest, i.e. the isotopic composition of the element must be different from that which would normally occur. This approach is illustrated in Figure 6.4. In this figure, spectrum (a) shows the

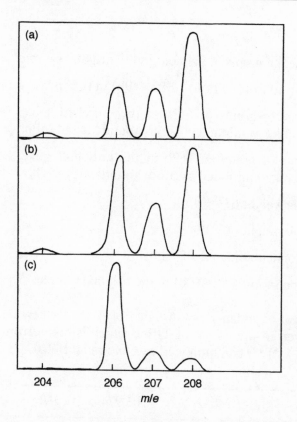

Figure 6.4 Isotopes of lead: (a) normal isotopic lead, (b) mixture of normal isotopic lead and enriched ^{206}Pb spike; (c) artificially enriched ^{206}Pb

naturally occurring isotopic composition for lead, i.e. the lead normally present in a sample. An artificially enriched isotopic standard, in which the abundance of ^{206}Pb has been increased by ca. 25% (spectrum (c)), is then added. The result is shown as spectrum (b). By measuring the isotopic ratio of the most abundant Pb isotope (^{208}Pb) to ^{206}Pb in the sample (^{208}Pb/^{206}Pb ~2.0), and comparing this with the value found for the sample plus spike (~1.0), an exact concentration of lead in the sample can be determined by using the following formula:

$$A = [(xB_2m_1/m_2) - B_1]/(z - zx/y) \tag{6.3}$$

where:

A = number of grams of element in the original sample;

x = measured isotope ratio ($^{208}Pb/^{206}Pb$) in the spiked sample;

B_1 = number of grams of ^{208}Pb in the enriched spike, i.e. mass of total spike multiplied by atom abundance of ^{208}Pb;

B_2 = number of grams of ^{206}Pb in the enriched spike, i.e. mass of total spike multiplied by atom abundance of ^{206}Pb;

m_1 = atomic weight of ^{208}Pb;

m_2 = atomic weight of ^{206}Pb;

z = fractional abundance of ^{208}Pb;

y = isotope ratio ($^{208}Pb/^{206}Pb$) in the original sample.

The fractional abundance (by weight) of ^{208}Pb is calculated by using the data given in the following table, which is taken from NIST 981 (see Section 1.5) for common lead (atomic weight = 207.215).

Isotope	203.973	205.974	206.976	207.977
Atom abundance (%)	1.425	24.144	22.083	52.347

The required parameter (z) is calculated as follows:

$$\frac{(207.977 \times 52.347)}{(203.973 \times 1.425) + (205.974 \times 24.144) + (206.976 \times 22.083) + (207.977 \times 52.347)}$$

$$= 0.5254.$$

It should be noted, however, that a quadrupole mass spectrometer does not normally measure the absolute isotopic composition for an element. It is likely, therefore, that the mass spectrometer will suffer from mass discrimination effects. If this is the case, it will be necessary to apply corrections to the measured isotopic ratios before carrying out the calculation using isotope dilution analysis.

SAQ 6.4

The concentration of lead in an aqueous sample (100 ml) was 100 ng ml^{-1}. This sample was spiked with an enriched ^{206}Pb standard (3.5 µg) by the addition of a 5 ml volume. After ICP-MS analysis, the following isotopic ratios were found:

System	^{208}Pb/^{206}Pb
Original sample	2.1681
Sample plus enriched ^{206}Pb spike	0.9521

The isotopic composition of normal lead (NIST 981) has been given earlier, while the corresponding composition of the enriched ^{206}Pb spike (NIST 983) is shown in the following table:

Isotope	203.93	205.974	206.976	207.977
Atom abundance (%)	0.0342	92.1497	6.5611	1.2550

By applying the principle of isotope dilution analysis, use the above data to confirm that the concentration of lead in the original sample is 100 ng ml^{-1}.

SAQ 6.4

Isotope dilution analysis can be applied to all elements with more than one isotope. In principal, therefore, it can be used for multi-element analysis. However, in practice it is normally reserved for single-element determinations due to the higher cost of the enriched stable isotopes and the increased time required for each analysis. It is worth noting that IDA does provide an ideal approach to internal standardisation, in which one of the element's own isotopes acts as the internal standard. The use of such a standard is known to improve the precision of the data. It is for this reason that IDA–ICP–MS is most commonly used by various bodies in the production and certification of reference materials, or in studies relating to nutrition, bioavailability and speciation.

6.5 MASS SPECTRAL INTERPRETATION

Table 6.5 Relative abundances of selected naturally occurring isotopes[a]

(a)

	Atomic mass unit (amu)												
Element	46	47	48	49	50	51	52	53	54	55	56	57	58
Ti	8.0	7.3	73.8	5.5	5.4								
V					0.2	99.8							
Cr					4.3		83.8	9.5	2.4				
Mn										100			
Fe											91.7	2.1	0.3
Ni									5.9		91.7	2.1	0.3
Ni													68.1

	amu												
Element	59	60	61	62	63	64	65	66	67	68	69	70	
Co	100												
Ni		26.2	1.1	3.6		0.9							
Cu					69.2		30.8						
Zn						48.6		27.9	4.1	18.8		0.6	

(b)

	amu												
Element[b]	82	83	84	85	86	87	88	89	90	91	92	93	94
(Kr)	11.6	11.5	57.0		17.3								
Rb				72.2		27.8							
Sr			0.5		9.9	7.0	82.6						
Y								100					
Zr									51.4	11.2	17.1		17.4
Nb												100	
Mo											14.8		9.3

Table 6.5 (*continued*)

	amu					
Element	95	96	97	98	99	100
Zr		2.8				
Mo	15.9	16.7	9.5	24.1		9.6
Ru		5.5		1.9	12.7	12.6
Rh						
Pd						
Ag						
Cd						

(c)

	amu						
Element	101	102	103	104	105	106	107
Ru	17.1	31.6		18.6			
Rh			100				
Pd		1.0		11.1	22.3	27.3	
Ag							51.8
Cd						0.9	

	amu												
Element	108	109	110	111	112	113	114	115	116	117	118	119	120
Pd	26.5		11.7										
Ag		48.2											
Cd	0.9		12.5	12.8	24.1	12.2	28.7		7.5				
In						4.3		95.7					
Sn					1.0		0.6	0.4	14.5	7.7	24.2	8.6	32.6
Te													0.09

Table 6.5 (*continued*)

	amu											
Element	121	122	123	124	125	126	127	128	129	130	131	132	133
Sn		4.6		5.8									
Sb	57.4		42.6										
Te		2.6	0.9	4.8	7.1	18.9		31.7		33.9			
I							100						
Xe				0.1		0.1		1.9	26.4	4.1	21.2	27.0	
Cs													100
Ba										0.10		0.10	

	amu							
Element	134	135	136	137	138	139	140	141	142
Xe	10.4		8.9						
Ba	2.4	6.6	7.9	11.2	71.7				
La					0.1	99.9			
Ce			0.2		0.3		88.4		11.1

(d)

	amu							
Element[b]	172	173	174	175	176	177	178	179	180	181	182
(Yb)	21.9	16.1	31.8		12.7						
Lu				97.4	2.6						
Hf			0.2		5.2	18.6	27.3	13.6	35.1		
Ta									0.0	99.9	
W										0.1	26.3

Table 6.5 (*continued*)

(e)

					amu								
Element[b]	200	201	202	203	204	205	206	207	208	209	210	211	212
(Hg)	23.1	13.2	29.9		6.9								
Tl				29.5		70.5							
Pb					1.4		24.1	22.1	52.4				
Bi										100			

(f)

						amu					
Element	230	231	232	233	234	235	236	237	238	239	240
Th			100								
U					0.0	0.7			99.3		

[a]Data taken from D.R. Lide, (Ed.), *CRC Handbook of Chemistry and Physics*, 73rd Edn, CRC Press, Boca Raton, FL, 1992/3
[b]Symbols in parentheses indicate elements that have incomplete percentage abundance totals (<100%)

SAQ 6.5   Using the data given in Table 6.5 interpret the mass spectra presented in Figure 6.5a

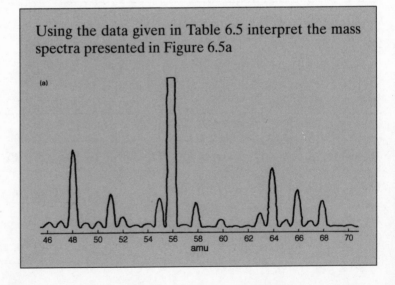

SAQ 6.5
(Contd)

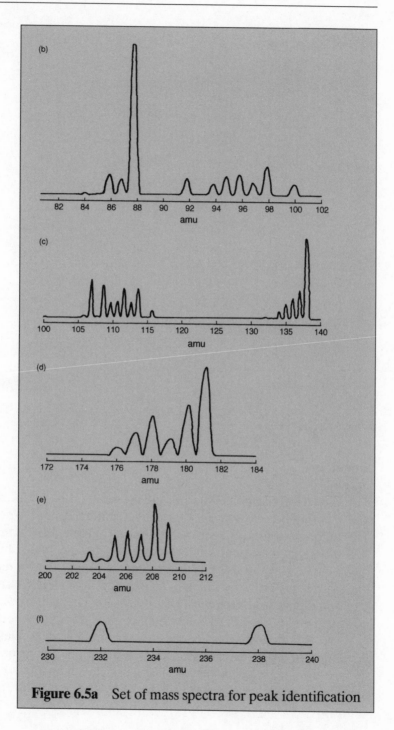

Figure 6.5a   Set of mass spectra for peak identification

SAQ 6.5

Summary

The inductively coupled plasma (ICP), as a result of its ease of use and flexibility, is now widely used as a source in mass spectrometry. The fundamentals and instrumental requirements for the technique of ICP inorganic mass spectrometry are described. A major limitation of this technique is the use of the quadrupole mass spectrometer with its limited resolution. This can lead to interferences. The causes of these, plus some potential remedies, are discussed. The use of solid samples, however, can offer certain advantages. An alternative source to the ICP, i.e. the glow discharge, can be used in this case. Details of the formation of a glow discharge and its applications are given. A unique feature of a mass spectrometer is its use in quantitative analysis for measuring isotope ratios. Making use of this ability, an example of the application of isotope dilution analysis is also presented.

Objectives

On completion of this chapter you should be able to:

- describe the operation of an inductively coupled plasma mass spectrometer;

- describe the nature of the interface;

- understand the principles of operation of a quadrupole mass spectrometer;

- appreciate the different methods of signal monitoring in mass spectrometry;

- understand the term, abundance sensitivity;

- appreciate why a high-resolution mass spectrometer may be required for certain applications;

- describe the operation of an electron multiplier tube;

- identify isobaric, molecular and matrix interferences in inductively coupled plasma mass spectrometry;

- carry out an isotope dilution (analysis) calculation;

- interpret inductively coupled plasma mass spectra.

7. Comparison of Tech

It is always difficult to compare techniques because of the diversity of information which they can produce. Therefore, in order to make any realistic comparison it is necessary to consider on what basis the techniques will be evaluated. An instrument manufacturer will frequently quote that a certain instrument has superior detection limits to another model, or that a particular modification will improve the sensitivity. While this information may well be correct, there is a difference between what is achievable in, for example, aqueous water samples and 'real' samples, as you should now be aware after studying this present text and, hopefully, also from some practical experience. Nevertheless, detection limits are frequently used as performance indicators. It is always advisable to note on what basis the detection limits are quoted (two or three times the standard deviation of the blank), and therefore compare the same approach. In addition to detection limits, various other parameters should be considered, and these are given below:

- precision;

- working linear range of instrument;

- instrument operation time (excluding sample preparation);

- capital cost of instrument;

- instrument operating costs;

- simultaneous or sequential multi-element analysis;

- ease of use of instrument;

- possibilities for automation;

likelihood of interferences and ability to remedy;

- flexibility of instrument, e.g. for the addition of a flow-injection assembly.

All of these parameters are important, but their degree of importance probably depends on whether you are a current user (in other words, you already have an instrument), or are looking to purchase one. Table 7 seeks to compare each of these parameters for the four general types of instrument covered in this present book. This table is by no means complete, and the results of your own personal experiences are probably as valid as those quoted.

7.1 CHARACTERISTIC CONCENTRATION

The characteristic concentration (or sensitivity) is a specific term which is used in atomic absorption spectroscopy for defining the magnitude of the absorbance signal which will be produced by a given concentration of the analyte. In flame atomic absorption spectroscopy, sensitivity is expressed as the concentration of an element (mg l^{-1}) which is required to produce a 1% absorption (0.0044 absorbance) signal; this can be expressed mathematically by the following relationship:

$$\text{sensitivity (mg } l^{-1}) = \frac{\text{concentration of standard solution (mg } l^{-1}) \times 0.0044}{\text{measured absorbance}}$$

$$(7.1)$$

Table 7 Comparison of the performance parameters of flame atomic absorption spectroscopy (FAAS), graphite furnace atomic absorption spectroscopy (GFAAS), inductively coupled plasma atomic emission spectroscopy (ICP-AES), and inductively coupled plasma mass spectrometry (ICP-MS)

Parameter	Technique			
	FAAS	GFAAS	ICP-AES	ICP-MS
(Typical) detection limits ng ml⁻¹	~ 100	~ 10–100	~ 100	~ 0.1–10
Precision[a]	Limited because it is not possible to use an internal standard; typically 1–2% RSD	Limited because it is not possible to use an internal standard; typically 1–3% RSD	With a simultaneous instrument it is possible to use an internal standard to improve precision; typically 0.1–1.0% RSD. Otherwise, similar to AAS techniques	Typically 0.1–1.0% RSD
Working linear range of instrument	Normally limited to up to one order of magnitude, e.g. 1–10 μl ml⁻¹ for a calibration	Normally limited to up to one order of magnitude, e.g. 0.1–1 μg ml⁻¹ for a calibration	Possible to work over may order of magnitude, e.g. 0.1–100 μg ml⁻¹	Possible to work over many orders of magnitude e.g. 0.001–1 μg ml⁻¹
Instrument operation time (excluding sample preparation)	Fast: sample introduced via nebuliser (allow 30 s), then signal acquired, followed by rinse-out time (up to 1 min)	Considered to be the slowest; a typical graphite furnace programme could take between 2 and 3 min	Fast: sample introduced via nebuliser (allow 30 s), then signal acquired followed by rinse-out time (up to 1 min)	Fast: sample introduced via nebuliser (allow 30 s), then signal acquired followed by rinse-out time (up to 1 min)

continued overleaf

Table 7 *(continued)*

Parameter	Technique			
	FAAS	GFAAS	ICP-AES	ICP-MS
Capital cost of instrument	Cheapest (~£10–15 K)	Intermediate (~£20–40 K)	High (~£50–100 K); depends whether simultaneous (higher cost) or sequential	Highest (~£150 K)
Instrument operating costs (excluding normal supply of electricity and water, as required)	Requires supply of acetylene gas, plus purchase of hollow-cathode lamps for the various elements to be analysed	Requires supply of nitrogen gas, plus purchase of hollow-cathode lamps for the various elements to be analysed. In addition, new graphite tubes are required on a regular basis	Requires large quantities of argon gas, plus periodic replacement of ICP torch	Requires large quantities of argon gas, plus periodic replacement of ICP torch. In addition, sample cones require regular replacement
Simultaneous or sequential multi-element analysis	(Normally) sequential	Sequential	Simultaneous and sequential	Simultaneous and sequential
Ease of use of instrument	'User friendly'	Requires some expertise to develop methods, plus initial training on the use of the graphite furnace	Operation is relatively straightforward when system is based on a nebuliser/spray chamber arrangement. Obviously, complexity is increased with other sample introduction devices. Ubiquitous software skills required for operation	Software training is essential for operation; comparatively straight-forward when based on a nebuliser/spray chamber arrangement (cf. ICP-AES). Obviously, complexity is increased with other sample introduction devices

Possibilities for automation	Autosampler can be easily added	Autosampler can be easily added	Autosampler can be easily added	Autosampler can be easily added
Likelihood of interferences and ability to remedy	Not normally a problem; interferences well characterised. Background correction is useful parameter	Interferences fairly well characterised. Background correction is required	Interferences well characterised. Background correction is required	Interferences well characterised but difficult to correct below 80 amu
Flexibility of instrument	Normally used with nebuliser/expansion chamber. Possible to add flow injection or chromatography assemblies with minimum alterations. Removal/alteration of burner assembly allows cold vapour or hydride generator to be used	Normally used for furnace work only. Difficult to interface other attachments	Normally used with nebuliser/spray chamber. Possible to add flow injection or chromatography assemblies with minimum alterations. Removal of nebuliser/spray chamber allows use of hydride generator, laser ablation or electrothermal vaporisation, as required	Normally used with nebuliser/spray chamber. Possible to add flow injection or chromatography assemblies with minimum alterations. Removal of nebuliser/spray chamber allows use of hydride generator, laser ablation or electrothermal vaporisation, as required

[a]RSD = relative standard deviation

8. Further Information

8.1 THE LABORATORY NOTEBOOK

When carrying out laboratory work it is essential to maintain an adequate record of your results. The following is intended as a general guide on how to keep a notebook.

The Record Book

First of all, it is important to *use a record book that has a permanent binding.* Loose-leaf, spiral bound or other temporarily bound books can allow for page removal, insertions and substitutions. In addition, the pages of the notebook should be numbered. The keeping of laboratory notebooks by entirely electronic methods is probably not advisable.

Ink Quality

When recording the results of experiments, do not use pencils or strange-coloured inks. Ensure that the ink is permanent, is not water- or solvent-reactive, and that it does not smear. It should also be light stable. The *use of black or blue ink is recommended* for normal use.

Entries

The entries in the book should be legible and factually complete.

For all experiments, it is important to describe, in as much detail as possible, *the actual procedure which is being carried out,* giving full descriptions of the experiment technique and all of the apparatus, with an accompanying sketch, if appropriate. *It is not necessary to repeat the information reported in the experimental script.*

In the case of atomic spectroscopy, full details of the apparatus used *must* be given (including operating conditions, types of instrument, supplier, etc.).

Drawings can be very important, so if in any doubt, include them. As a general guideline, there should be enough information in the record book to enable someone working in the field to duplicate the work.

Results and Observations

Record carefully all results and note all observations. In atomic spectroscopy, the record book should include all analytical data that have been obtained, plus *details of any calculations (with units)* that have been carried out.

Any graphs should be carefully fixed in the book by some permanent method (e.g. staples or pins) and appropriate reference made to them and their contents. *All axes should be labelled and the relevant units included.*

Any results subsequently added to the original data, for example, the results of analysis, should be entered on a separate page with a reference given to the original entry.

Never leave a page incomplete. Draw lines through unused pages or parts of pages.

Facts not Opinions

You should also record any novel concepts and ideas relating to the work. It is preferable not to express opinions in your notebook, as this could lead to misinterpretation. *The contents of the book should be limited to factual, quantitative and qualitative results.*

Do not use slang or abbreviations, unless the latter are defined, or well known. The notebook *must* be understandable to others reading it.

Supporting Information

Addition to the notebook of any supporting records (e.g. computer/integrator printouts) should not be made in an haphazard fashion. If such records cannot be added to the notebook itself, then reference to them should be consistent, and they should be stored elsewhere in an orderly, readily retrievable manner.

'Signing-off'

Ensure that *each page is signed and dated by the author.*

Errors

Errors and mistakes should not be erased or obliterated beyond recognition. Correction fluid should not be used. Simply crossing out an error so that the original entry is still legible should be adequate. *Never remove pages from the notebook.*

Safe Keeping

The notebook should be regarded as a form of 'legal' document. As such, it should be treated in a strictly controlled manner. *When completed, it should be stored in a safe place in either the laboratory or the office.*

8.2 ADDITIONAL READING MATERIAL

Books

General

R. Bock, *A Handbook of Decomposition Methods in Analytical Chemistry*, International Textbook Company, London, 1979.

H.M. Kingston and L.B. Jassie, *Introduction to Microwave Sample Preparation*, ACS Professional Reference Book, American Chemical Society, Washington, DC, 1988.

H.H. Willard, L.L. Meritt, Jr, J.A. Dean and F.A. Settle, Jr, *Instrumental Methods of Analysis*, 7th Edn, Wadsworth Publishing Company, Belmont, CA, 1988.

K.W. Busch and M.A. Busch, *Multielement Detection Systems for Spectrochemical Analysis*, Wiley, New York, 1990.

J.A.C. Broekaert, S. Gucer and F. Adams, *Metal Speciation in the Environment*, NATO ASI Series, Volume G23, Springer-Verlag, Berlin, 1990.

G.E. Baiuescu, P. Dumitrescu and P. Gh. Zugravescu, *Sampling*, Ellis Horwood, Chichester, 1991.

T.R. Crompton, *Analytical Instrumentation for the Water Industry*, Butterworth-Heinemann, Oxford, 1991.

D.A. Skoog and J.L. Leary, *Principles of Instrumental Analysis*, 4th Edn Saunders College Publishing, Orlando, FL, 1992.

T.R. Crompton, *The Analysis of Natural Waters, Volume 1: Complex-formation Preconcentration Techniques*, Oxford University Press, Oxford, 1993.

T.R. Crompton, *The Analysis of Natural Waters; Volume 2: Direct Preconcentration Techniques*, Oxford University Press, Oxford, 1993.

C. Vandecasteele and C.B. Block, *Modern Methods of Trace Element Determination*, Wiley, Chichester, 1993.

A.G. Howard and P.J. Statham, *Inorganic Trace Analysis. Philosophy and Practice*, Wiley, Chichester, 1993.

G. Christian, *Analytical Chemistry*, 5th Edn, Wiley, Chichester, 1994.

R.E. Seivers, *Selective Detectors: Environmental, Industrial and Biomedical Applications*, Wiley, Chichester, 1995.

S. Caroli, *Element Speciation in Bioinorganic Chemistry*, Wiley, Chichester, 1996.

B. Markert and I.H.L. Zittau, *Instrumental Element and Multi-element Analysis of Plant Samples. Methods and Applications*, Wiley, Chichester, 1996.

Quality Assurance

G. Kateman and L. Buydens, *Quality Control in Analytical Chemistry*, Wiley, Chichester, 1993.

M. Parkany (Ed.), *Quality Assurance for Analytical Laboratories*, Royal Society of Chemistry, Cambridge, 1994.

E. Pritchard, *Quality in the Analytical Chemistry Laboratory*, ACOL Series, Wiley, Chichester, 1995.

Statistics

J.C. Miller and J.N. Miller, *Statistics for Analytical Chemistry*, 2nd Edn, Ellis Horwood, Chichester, 1988.

Atomic Absorption and Atomic Emission Spectroscopy

L. Ebdon, *An Introduction to Atomic Absorption Spectroscopy*, Heyden and Son, London, 1982.

B. Welz, *Atomic Absorption Spectrometry*, VCH, Weinheim, 1985.

P.W.J.M. Boumans, *Inductively Coupled Plasma Emission Spectrometry*, Parts 1 and 2, Wiley, New York, 1987.

A. Montaser and D.W. Golightly, *Inductively Coupled Plasmas in Analytical Atomic Spectrometry*. VCH, New York, 1987.

J.D. Ingle and S.R. Crouch, *Spectrochemical Analysis*, Prentice-Hall, London, 1988.

M. Thompson and J.N. Walsh, *A Handbook of Inductively Coupled*

Plasma Spectrometry, 2nd Edn, Blackie Academic and Professional, Glasgow, 1989.

L. Moenke-Blankenburg, *Laser Microanalysis*, Wiley, New York, 1989.

G.L. Moore, *Introduction to Inductively Coupled Plasma Atomic Emission Spectroscopy*, Elsevier, Amsterdam, 1989.

R.M. Harrison and S. Rapsomanikis, *Environmental Analysis Using Chromatography Interfaced with Atomic Spectroscopy*, Ellis Horwood, Chichester, 1989.

J. Sneddon, *Sample Introduction in Atomic Spectroscopy*, Volume 4, Elsevier, Amsterdam, 1990.

L.H.J. Lajunen, *Spectrochemical Analysis by Atomic Absorption and Emission*, The Royal Society of Chemistry, Cambridge, 1992.

J. Sneddon, *Advances in Atomic Spectroscopy*, JAI Press, Greenwich, CT, 1992.

K. Slickers, *Automatic Atomic Emission Spectroscopy*, 2nd Edn, Bruhlsche Universitatsdruckerei, Giessen, 1993.

M.S. Cresser, *Flame Spectrometry in Environmental Chemical Analysis: A Practical Approach*, The Royal Society of Chemistry, Cambridge, 1995.

Z. Fang, *Flow Injection Atomic Absorption Spectrometry*, Wiley, Chichester, 1995.

J. Dedina and D.I. Tsalev, *Hydride Generation Atomic Absorption Spectrometry*, Wiley, Chichester, 1995.

Inductively Coupled Plasma Mass Spectrometry

F. Adams, R. Gijbels and R. van Grieken, *Inorganic Mass Spectrometry*, Wiley, New York, 1988.

A.R. Date and A.L. Gray, *Applications of Inductively Coupled Plasma Mass Spectrometry*, Blackie Academic and Professional, Glasgow, 1989.

K.E. Jarvis, A.L. Gray, I. Jarvis and J.G. Williams, *Plasma Source Mass Spectrometry*, The Royal Society of Chemistry, Cambridge, 1990.

G. Holland and A.N. Eaton, *Applications of Plasma Source Mass Spectrometry*, The Royal Society of Chemistry, Cambridge, 1991.

K.E. Jarvis, A.L. Gray and R.S. Houk, *Handbook of Inductively Coupled Plasma Mass Spectrometry*, Blackie Academic and Professional, Glasgow, 1992.

E.H. Evans, J.J. Giglio, T.M. Castillano and J.A. Caruso, *Inductively Coupled and Microwave Induced Plasma Sources for Mass Spectrometry*, The Royal Society of Chemistry, Cambridge, 1995.

Glow Discharges

R.K. Marcus, *Glow Discharge Spectroscopies*, Plenum, New York, 1993.

Journals

Regular articles on the latest advances in atomic spectroscopy appear in the following journals:

Journal of Analytical Atomic Spectrometry

The Analyst

Analytical Communications

Analytical Chemistry

Analytica Chimica Acta

Spectrochimica Acta, Part B

Applied Spectroscopy

Fresenius Zeitschrift für Analytische Chemie

Talanta

In addition, the Journal of Analytical Atomic Spectrometry features regular reviews on applications and techniques in a secton entitled Atomic Spectrometry Updates.

Self-assessment Questions and Responses

SAQ 1.2

Which of the sets of results shown in Figure 1.2a would you describe as accurate and which as precise? It is possible to come up with suitable answers (i.e. high accuracy, low accuracy, high precision, or low precision) for each bull's-eye.

Response

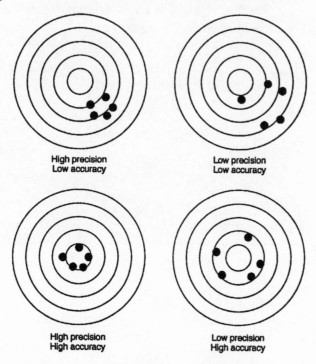

High precision
Low accuracy

Low precision
Low accuracy

High precision
High accuracy

Low precision
High accuracy

Figure 1.2b Identification of the sets of results

SAQ 1.3

(a) A River Tyne water sample was analysed by inductively coupled plasma (ICP) atomic emission spectroscopy (AES) for Ni. The sample was nebulised directly and the signal obtained was 15.6 mV. A calibration plot was generated by using 0, 1, 5, 10, and 50 ppm solutions, which gave the following responses, i.e. 0, 2, 9, 18 and 93 mV. What is the concentration of Ni in the original River Tyne water sample?

(b) A coastal sea water sample was analysed by ICP-AES for Cu. The sample was diluted by placing 10 ml in a 100 ml volumetric flask and adding 90 ml of distilled water. A calibration plot was generated by diluting a 1000 ppm stock solution. 10 ml of the stock solution were then placed in a 100 ml volumetric flask and made up to the mark with distilled water (working solution). This solution was diluted in series as follows:

Flask	Cu working solution (ml)	Water (ml)	Total volume (ml)
1	0	100	100
2	1	99	100
3	2	98	100
4	3	97	100
5	5	95	100

SAQ 1.3
(Contd)

The signals obtained were as follows:

Flask	Signal (mV)
1	0
2	150
3	290
4	435
5	730
Diluted sample	490

What is the concentration of Cu in the original coastal sea water sample?

(c) 0.5020 g of a steel sample was digested in concentrated acid and then transferred to a volumetric flask (100 ml) and made up to the mark with distilled water. The sample was then diluted 10 times. The diluted sample was then analysed for Pb as follows:

Flask	Volume of 100 ppm Pb solution (ml)	Digested and diluted sample (ml)	Volume of water (ml)	Total volume (ml)
1	0	20	80	100
2	1	20	79	100
3	2	20	78	100
4	3	20	77	100
5	5	20	75	100
6	7	20	73	100

→

SAQ 1.3
(Contd)

After analysis, the following results were obtained:

Flask	Signal (mV)
1	29
2	37
3	44
4	52
5	68
6	83

Calculate the concentration of Pb in the original sample in units of $\mu g\ g^{-1}$ and wt%.

(d) A sample of soil was accurately weighed (0.5250 g) into a microwave vessel and 9 ml of concentrated HNO_3 and 3 ml of concentrated HF were added. After heating for 20 min in a microwave oven the sample was allowed to cool. The contents of the vessel were quantitatively transferred to a 100 ml volumetric flask and analysed for Cu and Ni by using an ICP-AES instrument. Calibration of the instrument was carried out by using suitable dilutions of 1000 $\mu g\ ml^{-1}$ Cu and Ni stock solutions.

Initially, 10 ml of each stock solution were diluted by addition to a 100 ml volumetric flask and the appropriate amount of dilute acid added (diluted stock solution). Then, the (diluted stock) solutions were diluted further, according to the following procedure, to obtain a series of calibration solutions.:

SAQ 1.3
(Contd)

Flask	Diluted stock solution/ volume added to a 100 ml volumetric flask (ml)
1	0
2	1
3	5
4	10
5	20

Using the calibration solutions, the digested sample was analysed, with the following results being obtained:

Flask	Cu ICP-AES signal (mV)	Ni ICP-AES signal (mV)
1	0	0
2	7	15
3	28	45
4	53	104
5	101	205
Digested sample solution	82	75

Calculate the concentrations (in mg kg^{-1}) of nickel and copper in the original soil sample.

(e) When a sample of River Tyne water was analysed for Pb by direct aspiration into an atomic absorption spectrometer equipped with an air–acetylene flame an absorbance reading of 0.25 was obtained. A calibration plot was prepared by diluting a standard 1000 μg ml^{-1}

→

SAQ 1.3
(Contd)

solution to $10 \, \mu g \, ml^{-1}$ and then taking the following volumes: 0, 10, 20, 30, and 50 ml. Each aliquot was then diluted to 100 ml with water and analysed by flame atomic absorption spectroscopy (FAAS). The following results were obtained:

Diluted stock solution $10 \, \mu g \, ml{-1} \, Pb$	Absorbance reading
0	0.000
10	0.082
20	0.162
30	0.245
50	0.410

Calculate the concentration (in $\mu g \, ml^{-1}$) of Pb in the sample.

Response

(a) The concentration of Ni in the original sample is 8.5 ppm.

(b) From your graph, you should obtain a value of 3.4 ppm. After taking into account the dilution factor, the concentration of Cu in the sea water sample is 34 ppm.

(c) From your graph, you should obtain a value of 3.8 ppm. After taking into account the dilution factor, the concentration of Pb in the steel sample is $37\,850 \, \mu g \, g^{-1}$ or 3.79 wt%.

(d) From your graph, you should obtain values of 16 (Cu) and 7.4 (Ni) ppm. After taking into account the dilution factors, the concentrations of Cu and Ni in the soil sample are 3048 and $1410 \, mg \, kg^{-1}$, respectively.

(e) The concentration of Pb in the water sample is $3.0 \, \mu g \, ml^{-1}$.

SAQ 2.1   Is it possible to separate tin (Sn) from lead (Pb) by forming an APDC complex?

Response

By changing the pH of the solution it is potentially possible to achieve selectivity between Sn and Pb. Tin can only form an APDC complex in the pH range 2–8, while Pb forms a complex between pH 2 to 14 (see Table 2.1a). Therefore, if the Pb complex was formed at pH values > 8, then no similar complex for Sn should be formed. In this way, selectivity between Pb and Sn should be possible.

SAQ 2.2   Why should HF be so effective for the digestion of silica?

Response

Hydrofluoric acid is the reagent used for dissolving silica-based materials. The silicates are converted to a more volatile species in solution:

$$SiO_2 + 6HF = H_2(SiF_6) + 2H_2O$$

SAQ 3.2a	If the frequency of electromagnetic radiation is 5×10^{14} Hz, what is the wavelength of this radiation and in which spectral region does it occur?

Response

In order to calculate the wavelength, we need to rearrange equation (3.1) as follows:

$$\lambda = c/f$$

By insertion of the appropriate values into this equation, we obtain the following:

$$\frac{3 \times 10^8 \, \text{m s}^{-1}}{5 \times 10^{14} \, \text{s}^{-1}} = 6 \times 10^{-7} \, \text{m}$$

While the value from the calculation is 6×10^{-7} m, it is normal to represent the value as 600 nm (or 0.000 000 600 m). This wavelength, 600 nm, occurs in the visible region of the electromagnetic spectrum.

SAQ 3.2b	Calculate the energy in joules of one photon of the radiation derived in SAQ 3.2a.

Response

$$E = 6.626 \times 10^{-34} \, \text{J s} \times 5 \times 10^{14} \, \text{s}^{-1} = 3.31 \times 10^{-19} \, \text{J}$$

or

$$E = \frac{6.626 \times 10^{-34} \, \text{J s} \times 3 \times 10^8 \, \text{m s}^{-1}}{600 \times 10^{-9} \, \text{m}} = 3.31 \times 10^{-19} \, \text{J}$$

SAQ 3.4 Confirm that the energy difference between the 3p and 3s levels in Figure 3.4b corresponds to the expected wavelength (note: $1\,\text{eV} = 1.602 \times 10^{-19}\,\text{J}$).

Response

First, convert $2.107\,\text{eV}$ to J, i.e.

$$2.107 \times 1.602 \times 10^{-19}\,\text{J} = 3.375 \times 10^{-19}\,\text{J}$$

Recall that $E = hc/\lambda$, and therefore:
$\lambda = hc/E$

$$\lambda = \frac{6.626 \times 10^{-34}\,\text{J}\,\text{s} \times 3 \times 10^{8}\,\text{m}\,\text{s}^{-1}}{3.375 \times 10^{-19}\,\text{J}}$$

$\lambda = 5.889(8) \times 10^{-7}\,\text{m}$, or $589\,\text{nm}$

SAQ 3.5 A typical temperature of an inductively coupled plasma is $7000\,\text{K}$ (see Section 5.2.1). What is the value of the N_1/N_0 population ratio?

Response

In an ICP, 6×10^{-2} (or 6%) of the atoms are in the excited state. Such a plasma is the most common source for atomic emission spectroscopy.

SAQ 3.6a For a sodium emission line of wavelength $589\,\text{nm}$ and an excited lifetime of $2.5 \times 10^{-9}\,\text{s}$, what is the natural linewidth?

Response

The value of the linewidth ($\Delta\lambda_{\text{N}}$) is $0.000\,46\,\text{nm}$.

SAQ 3.6b For a sodium atom in a flame at a temperature of 2500 K calculate the Doppler linewidth for the 589 nm spectral line (note that the sodium atomic mass is 23 g mol^{-1}, but in SI units (Système Internationale d'Unités) the value is 23 × 10^{-3}/kg mol^{-1}).

Response

$$\Delta\lambda_D = (2 \times 589 \times 10^{-9}\,m /3 \times 10^8\,m\,s^{-1})$$

$$\sqrt{(2 \times 8.314\,J\,K^{-1}\,mol^{-1} \times 2500\,K / 23 \times 10^{-3}\,kg\,mol^{-1})}$$

which becomes:

$$\Delta\lambda_D = (3.927 \times 10^{-15}\,s)\sqrt{1\,807\,391}\ J\,kg^{-1})$$

Recall that $1\,J = 1\,kg\,m^2\,s^{-2}$, so:

$$\Delta\lambda_D = (3.927 \times 10^{-15}\,s)\sqrt{(1807\,391\,kg\,m^2\,s^{-2}\,kg^{-1})}$$

By taking the square root, we obtain:

$$\Delta\lambda_D = (3.927 \times 10^{-15}\,s)\,(1344\,m^{2/2}\,s^{-2/2})$$

which simplifies to give the following:

$$\Delta\lambda_D = (3.927 \times 10^{-15}\,s)\,(1344\,m\,s^{-1})$$

We finally obtain:

$$\Delta\lambda_D = 5.28 \times 10^{-12}\,m, \text{ i.e. } 5.3\,pm \text{ or } 0.005\,3\,nm$$

SAQ 4.3 Write a simple chemical equation to explain the sputtering process that occurs in the hollow-cathode lamp.

Response

$$M(s) \xrightarrow{\text{Ar}^+(g)} M(g)$$

Upon liberation, the metal atoms can further collide with the fill-gas ions and electrons, thus causing excitation of the metal:

$$M(g) \xrightarrow{e^-, \text{Ar}^+} M^*(g)$$

The excited metal atoms return to the ground state and emit characteristic wavelengths of the metal itself and also the fill gas. Therefore, the hollow-cathode lamp is an emission source.

SAQ 4.4a Would you think it necessary to operate with a gas flow during the atomisation step?

Response

There is normally no flow of gas during the atomisation step so as to maximise the residence time of the metal atoms in the tube, thus increasing the sensitivity of the technique.

SAQ 4.4b   Why does the atom cell require heating in hydride generation?

Response

The nature of the AAS technique requires the analyte to be present as atoms. Hydride generation produces volatile molecular species, e.g. AsH_3, which require dissociation prior to detection.

SAQ 4.5   Compare the pneumatic nebulisers used for flame atomic absorption and inductively coupled plasma atomic emission spectroscopy.

Response

While both nebulisers may look the same when represented schematically, their physical appearance is quite different. The pneumatic nebuliser for AES is constructed of glass, while for AAS the nebuliser is made of steel. In addition, the tolerance in manufacture for the AES type is much higher than that required for AAS. For further details, the reader is referred to the appropriate sections in this book.

SAQ 4.7   Why should the presence of molecular species be a problem in AAS?

Response

The temperatures achieved in the flame or the graphite furnace are not sufficient to dissociate all of the molecular species that are present into atoms. Various molecular components can be produced from gaseous species, e.g. the surrounding air, or the fuel-oxidant mixture, or they can be sample-derived, e.g. from the acids used to dissolve the solid sample.

SAQ 4.8a   Can you think of any remedies to alleviate chemical interferences?

Response

Several remedies are possible, and may include the following: (i) the use of a hotter flame, e.g. nitrous oxide–acetylene, which would dissociate the thermally stable components that are formed; (ii) the addition of a releasing agent, e.g. Sr or La, which preferentially reacts with phosphate to form a thermally stable compound (see Figure 4.8a); (iii) the addition of a protective chelating agent, e.g. EDTA, which allows preferential complexation and the formation of relatively thermally unstable alkaline earth metal–EDTA complexes.

SAQ 4.8b   Explain how you would prepare a standard additions plot for the analysis of calcium in sea water.

Response

In the standard additions method, all of the working (standard) solutions contain the same volume of sample, i.e. the sea water, while the concentration of the calcium solution is varied. Each of these working solutions is introduced into the spectrometer and the response recorded. However, as the graph produced by plotting the signal response (e.g. the absorbance in atomic absorption spectroscopy) against the analyte concentration does not pass through zero for either axis, extrapolation is required until the line crosses the x-axis. By maintaining a constant concentration on the x-axis, the unknown sample concentration can be determined (see Figure 1.3b). It is essential that this graph is linear over its entire length, otherwise considerable errors can be introduced.

The following table illustrates the various operations required in the preparation of solutions for the method of standard additions:

Flask	Volume of sea water (ml)	Concentration of calcium ($\mu g\,ml^{-1}$)	Total liquid volume of volumetric flask[a] (ml)
1	10	0	100
2	10	1	100
3	10	3	100
4	10	5	100
5	10	10	100

[a]Assuming a capacity of 100 ml

SAQ 4.8c

Can you suggest any elements that form carbides?

Response

The best known is probably tungsten, but various other elements can also form carbides, e.g. barium, vanadium, molybdenum and tantalum.

SAQ 5.1   The introduction of a sodium salt into a Bunsen
burner is characterised by the colour of the flame.
What colour would you expect the flame to be?

Response

Sodium, with its strong doublet at about 589 nm, is characterised by an
intense yellow colour.

SAQ 5.2a   Can you identify any potential difficulty in viewing
the luminous plasma side-on? Can you suggest an
alternative viewing position?

Response

Side-on viewing results in a significant background emission being
observed in addition to the element of interest (see Figure 5.2c). It has
been suggested by several research groups that end-on or axial
viewing of the central injector channel of the plasma torch will lead to
greater sensitivity and an improved signal-to-background ratio. This
latter feature has recently been developed commercially.

SAQ 5.2b   Do you think that the position of the cathode
directly above the two anodes offers any advantage
for the DCP?

Response

The configuration of the electrodes promotes stabilisation in the
discharge plume.

SAQ 5.2c   Would the ICP be a good source for the analysis of non-metals?

Response

No — the ionisation energy of Ar (15.76 eV) is considerably less than that of He (24.59 eV), so that the ionisation of metals (with typical ionisation energies for the first-row transition metals ranging from 6.56 eV for scandium to 9.39 eV for zinc) is easier than the ionisation of non-metals (e.g. Cl 12.97, N 14.53, P 10.49, and S 10.36 eV).

SAQ 5.3a   Sample introduction in AES requires the conversion of the sample into a (dry or wet) aerosol. Why should the use of an aerosol be an effective means of introducing the sample?

Response

Generation of an aerosol prior to introduction into the plasma conserves the plasma's temperature. If a large amount of sample was introduced directly into the plasma this would cause cooling. Remember that a 'cooler' source, e.g. an air–acetylene flame, can be a major source of chemical interferences (See Section 4.8.1).

SAQ 5.3b   The laser most commonly used for ablation is the Nd–YAG type. What does Nd–YAG stand for?

Response

Neodymium–yttrium aluminium garnet.

SAQ 5.3c	Design a flow injection system for the on-line mixing of two carrier streams. Can you think of a possible way of calibrating a system by using such a methodology?

Response

Figure 5.3h (a) shows a method for mixing two carrier streams. In principle, it should be possible to have a system in which the sample carrier stream is mixed with a carrier stream into which different standard solutions could be introduced. This would provide a method for generating a standard additions calibration plot on-line.

SAQ 5.4a	Would you envisage any difficulties in operating a spectrometer below 190 nm?

Response

Below 190 nm, purged optics (using N_2) and a vacuum spectrometer are required due to the absorption of oxygen. The typical operation of most spectrometers, however, is between 190 and 450 nm.

SAQ 5.4b	Calculate the wavelength of a spectral emission line, of the first order, with a groove density of 1200 lines per mm and an angle (ϕ) of 20°.

Response

A groove density of 1200 lines per mm is equivalent to one groove every 0.833×10^{-6} m (or 0.833 μm); therefore, $d = 0.833$ μm. The wavelength would then be 284 nm. At what wavelength would this same emission line occur for the second order? The answer is 142 nm.

SAQ 5.4c   After considering the spectral information given in Figure 5.4i, which do you think would be the best photomultiplier tube for use in a monochromator?

Response

The photomultiplier tube that gives the widest spectral wavelength coverage, i.e. 190–800 nm, is trialkali (Na–K–Sb–Cs).

SAQ 6.2a What type of mass spectrometer do you think could be used in ICP-MS?

Response

Any type of mass spectrometer could potentially be used, e.g. ion trap, time-of-flight, etc. However, the first to be commercially developed was the quadrupole mass spectrometry system. Future developments in ICP-MS technology will undoubtedly see further systems becoming commercially available.

SAQ 6.2b Where do the neutralised ions end up in the system?

Response

As the mass spectrometer is a vacuum system that operates by using appropriate pumps (rotary, turbomolecular, etc.) it is the oil contained within these pumps which acts as a reservoir for the trace metals. Periodic replacement of the oil is therefore required.

SAQ 6.2c What is the difference between sequential and simultaneous multi-element analysis? What analogy does this provide in ICP optical spectroscopy?

Response

Sequential multi-element analysis is where only one element is determined at any one time, although the solutions themselves may contain many elements for analysis. In simultaneous multi-element analysis, all of the elements present in the solution can be determined at the same time. In ICP optical spectroscopy, sequential multi-element analysis would be carried out by using a monochromator (Czerny–Turner configuration), whereas in simultaneous multi-element analysis a polychromator (Paschen–Runge configuration) would be employed.

SAQ 6.2d   Compare the operation of the electron multiplier tube with that of the photomultiplier tube.

Response

The reader should consult Sections 5.4.3 and 6.2.5, and make appropriate notes.

SAQ 6.2e   Can you identify a situation, in which an alternative mass could be used, for the determination of zinc in a nickel alloy?

Response

In this case, Zn atomic mass 66 is suitable (Zn, 27.8% abundance), in preference to Zn mass 64 at which Ni (0.95% abundance) also occurs.

SAQ 6.2f   Can you also identify a similar situation for titanium, in which the only available alternative mass is one with an inherently low percentage abundance?

Response

In this situation, Ti atomic mass 49 is suitable (Ti, 5.5% abundance), in preference to Ti mass 48 at which Ca (0.19% abundance) also occurs.

SAQ 6.2g   Can you identify a situation for selenium in which no alternative mass is available?

Response

It may appear that selenium (atomic masses of 78, 80 and 82) could have an isobaric interference from krypton. As it is most unlikely that any krypton would be determined in the argon gas, this is not a problem. As we shall see later, a major problem does occur at these mass numbers (78, 80 and 82) but is of an alternative type in this case.

SAQ 6.2h   Barium has atomic mass values of 130, 132, 134, 135, 136, 137, and 138. By using this information, can you identify the atomic masses at which Ba^{2+} species will occur?

Response

Ba^{2+} ions shall occur at half the masses of the singly charged parent ions. In order of importance, according to the magnitude of their effect (i.e. the greatest first), the Ba^{2+} ions will occur at atomic masses of 69, 68, 67, 66, and 65.

SAQ 6.3a   How does the operation of a glow discharge source compare with that of a hollow-cathode lamp (as used in atomic absorption spectroscopy)?

Response

Operation of the glow discharge source and the hollow-cathode lamp is essentially identical. The only difference of any note is that the glow discharge source contains a replaceable cathode, which the hollow cathode lamp does not have. Certain cosmetic differences are also apparent.

SAQ 6.3b   Why should changing from an argon to an helium fill-gas affect sputtering?

Response

The masses of the bombarding atoms are heavier in the case of argon so these atoms have more momentum when they strike the surface.

SAQ 6.3c   Explain why the doubly charged ion species, $^{56}Fe^{2+}$, occurs at 28 atomic mass units (amu).

Response

Mass spectrometry separates on the basis of mass/charge ratio, i.e. $56/2 = 28$.

SAQ 6.4

The concentration of lead in an aqueous sample (100 ml) was 100 ng ml^{-1}. This sample was spiked with an enriched ^{206}Pb standard (3.5 µg) by the addition of a 5 ml volume. After ICP-MS analysis, the following isotopic ratios were found:

System	^{208}Pb/^{206}Pb
Original sample	2.1681
Sample plus enriched ^{206}Pb spike	0.9521

The isotopic composition of normal lead (NIST 981) has been given earlier, while the corresponding composition of the enriched ^{206}Pb spike (NIST 983) is shown in the following table:

Isotope	203.93	205.974	206.976	207.977
Atom abundance (%)	0.0342	92.1497	6.5611	1.2550

By applying the principle of isotope dilution analysis, use the above data to confirm that the concentration of lead in the original sample is 100 ng ml^{-1}.

Response

Isotope dilution analysis, using isotopic ratios, is an effective method for the determination of the concentration of a particular element. In this present case, you will first need to calculate values for the abundance of both ^{206}Pb and ^{208}Pb by using the tables provided (NIST 983 and NIST 981, respectively). The results obtained, along with the other data provided, are then substituted into equation (6.3) to obtain the required concentration of lead in the original sample. If you have carried out the calculation correctly you should obtain a value of 98.8 ng ml^{-1}.

SAQ 6.5   Using the data given in Table 6.5, interpret the mass spectra presented in Figure 6.5a.

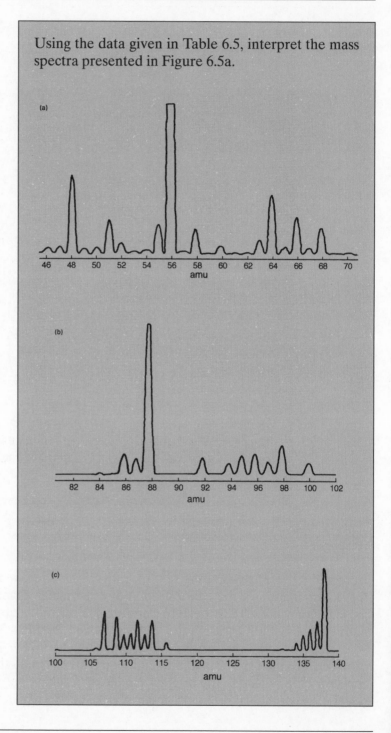

SAQ 6.5

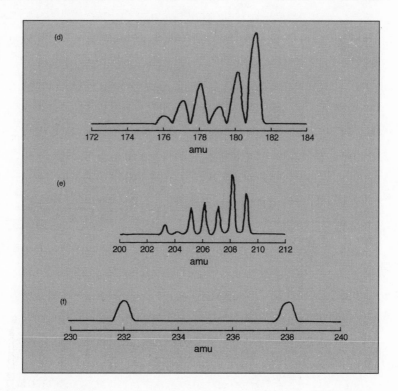

Response

Assignment of the peaks to various elements over the range from 46 to 240 amu is given in the annotated spectra presented in Figure 6.5b.

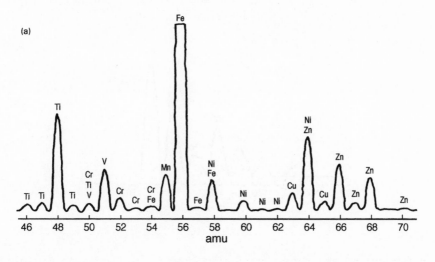

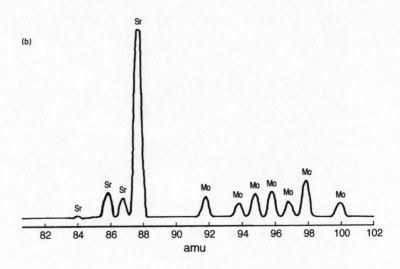

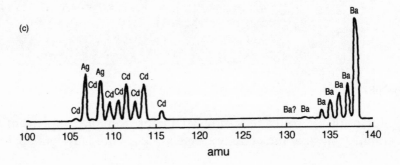

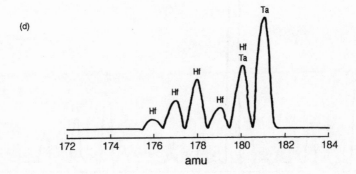

(Contd)

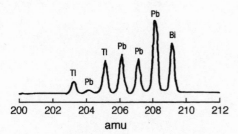

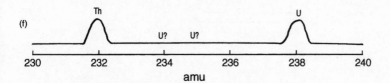

Figure 6.5b Annotated mass spectra showing assignment of peaks to various elements

Units of Measurement

For historical reasons a number of different units of measurement have evolved to express a quantity of the same thing. In the 1960s, many international scientific bodies recommended the standardisation of names and symbols and the adoption universally of a coherent set of units — the SI units (Système Internationale d'Unités) — based on the definition of seven basic units: metre (m), kilogram (kg), second (s), ampere (A), kelvin (K), mole (mol), and candela (cd).

The earlier literature references and some of the older text books naturally use the older units. Even today, many practising scientists have still not adopted SI units as their working units. It is, therefore, necessary to know of the older units and to be able to interconvert these with the SI units.

In this series of texts, SI units are used as standard practice. However, in areas of activity where their use has not become general practice, for example biologically based laboratories, the earlier defined units are used. This is explained in the study guide to each unit.

Table 1 shows some symbols and abbreviations commonly used in analytical chemistry, while Table 2 shows some of the alternative methods for expressing the values of physical quantities and their relationship to the values in SI units. In addition, Table 3 lists prefixes for SI units and Table 4 shows the recommended values of a selection of physical constants.

Further details and definitions of other units may be found in I. Mills, T. Cvitaš, K. Homann, N. Kallay and K. Kuchitsu, *Quantities, Units and Symbols in Physical Chemistry*, 2nd Edn, Blackwell Science, 1993.

Table 1   Symbols and abbreviations commonly used in analytical
chemistry

Å	angstrom
$A_r(X)$	relative atomic mass of X
A	ampere
E or U	energy
G	Gibbs free energy (function)
H	enthalpy
I (or i)	electric current
J	joule
K	kelvin ($= 273.15 + t\,(°C)$)
K	equilibrium constant (with subscripts p, c, etc.)
K_a, K_b	acid and base ionisation constants
$M_r(X)$	relative molecular mass of X
N	newton (SI unit of force)
P	total pressure
s	standard deviation
T	temperature (K)
V	volume
V	volt ($J\ A^{-1}\ s^{-1}$)
a, a(A)	activity, activity of A
c	concentration (mol dm^{-3})
e	electron
g	gram
s	second
t	temperature (°C)
b.p.	boiling point
f.p.	freezing point
m.p.	melting point
~	approximately equal to
<	less than
>	greater than
e, exp(x)	exponential of x
ln x	natural logarithm of x; $\ln x = 2.303 \log x$
log x	common logarithm of x to base 10

Table 2   Summary of alternative methods of expressing physical quantities

(1)   Mass (SI unit: kg)

$$g = 10^{-3} \, kg$$
$$mg = 10^{-3} \, g = 10^{-6} \, kg$$
$$\mu g = 10^{-6} \, g = 10^{-9} \, kg$$

(2)   Length (SI unit: m)

$$cm = 10^{-2} \, m$$
$$\text{Å} = 10^{-10} \, m$$
$$nm = 10^{-9} \, m = 10 \, \text{Å}$$
$$pm = 10^{-12} \, m = 10^{-2} \, \text{Å}$$

(3)   Volume (SI unit: m^3)

$$l = dm^3 = 10^{-3} \, m^3$$
$$ml = cm^3 = 10^{-6} \, m^3$$
$$\mu l = 10^{-3} \, cm^3$$

(4)   Concentration (SI unit: $mol \, m^{-3}$)

$$M = mol \, l^{-1} = mol \, dm^{-3} = 10^3 \, mol \, m^{-3}$$
$$mg \, l^{-1} = \mu g \, cm^{-3} = ppm = 10^{-3} \, g \, dm^{-3}$$
$$\mu g \, g^{-1} = ppm = 10^{-6} \, g \, g^{-1}$$
$$ng \, cm^{-3} = ppb = 10^{-6} \, g \, dm^{-3}$$
$$pg \, g^{-1} = ppt = 10^{-12} \, g \, g^{-1}$$
$$mg\% = 10^{-2} \, g \, dm^{-3}$$
$$\mu g\% = 10^{-5} \, g \, dm^{-3}$$

(5)   Pressure (SI unit: $N \, m^{-2} = kg \, m^{-1} s^{-2}$)

$$Pa = N \, m^{-2}$$
$$atm = 101\,325 \, N \, m^{-2}$$
$$bar = 10^5 \, N \, m^{-2}$$
$$torr = mmHg = 133.322 \, N \, m^{-2}$$

(6)   Energy (SI unit: $J = kg \, m^2 s^{-2}$)

$$cal = 4.184 \, J$$
$$erg = 10^{-7} \, J$$
$$eV = 1.602 \times 10^{-19} \, J$$

Table 3 Prefixes for SI units

Fraction	Prefix	Symbol
10^{-1}	deci	d
10^{-2}	centi	c
10^{-3}	milli	m
10^{-6}	micro	μ
10^{-9}	nano	n
10^{-12}	pico	p
10^{-15}	femto	f
10^{-18}	atto	a

Multiple	Prefix	Symbol
10	deca	da
10^2	hecto	h
10^3	kilo	k
10^6	mega	M
10^9	giga	G
10^{12}	tera	T
10^{15}	peta	P
10^{18}	exa	E

Table 4   Recommended values of physical constants

Constant	Symbol	Value
acceleration due to gravity	g	$9.81 \, \text{m s}^{-2}$
Avogadro constant	N_A	$6.022\,14 \times 10^{23} \, \text{mol}^{-1}$
Boltzmann constant	k	$1.380\,66 \times 10^{-23} \, \text{J K}^{-1}$
charge-to-mass ratio	e/m	$1.758\,796 \times 10^{11} \, \text{C kg}^{-1}$
electronic charge	e	$1.602\,18 \times 10^{-19} \, \text{C}$
Faraday constant	F	$9.648\,46 \times 10^{4} \, \text{C mol}^{-1}$
gas constant	R	$8.314 \, \text{J K}^{-1} \text{mol}^{-1}$
ice-point temperature	T_{ice}	$273.150 \, \text{K}$[a]
molar volume of ideal gas (stp)	V_m	$2.241\,38 \times 10^{-2} \, \text{m}^3 \text{mol}^{-1}$
permittivity of a vacuum	ε_0	$8.854\,188 \times 10^{-12}$ $\text{kg}^{-1} \text{m}^{-3} \text{s}^4 \text{A}^2 \, (\text{F m}^{-1})$
Planck constant	h	$6.626\,08 \times 10^{-34} \, \text{J s}$
standard atmosphere (pressure)	p	$101\,325 \, \text{N m}^{-2}$[a]
atomic mass constant	m_u	$1.660\,54 \times 10^{-27} \, \text{kg}$
speed of light in a vacuum	c	$2.997\,925 \times 10^{8} \, \text{m s}^{-1}$

[a]Exact value

Index